THE IMPACT OF THE GENE

COLIN TUDGE

𝕎 HILL **AND** WANG

A division of Farrar, Straus and Giroux

New York

THE **IMPACT**

OF THE **GENE**

—

From **Mendel's Peas**
to **Designer Babies**

Hill and Wang
A division of Farrar, Straus and Giroux
19 Union Square West, New York 10003

Copyright © 2000 by Colin Tudge
All rights reserved
Distributed in Canada by Douglas & McIntyre Ltd.
Printed in the United States of America
Originally published in 2000 by Jonathan Cape, United Kingdom, as
In Mendel's Footnotes
Published in the United States by Hill and Wang
First American edition, 2001

Library of Congress Cataloging-in-Publication Data
Tudge, Colin,
    The impact of the gene : from Mendel's peas to designer babies / Colin
Tudge— 1st ed.
        p.   cm.
    Includes bibliographical references and index.
    ISBN 0-374-17523-3 (hc. : alk. paper)
    1. Genetics—Popular works.

QH437.T833 2001
576.5—dc21   00-067306

Designed by Jonathan D. Lippincott

Contents

THE IMPACT OF THE GENE

The Future of Humankind and the Legacy of Mendel

Heredity matters. It is perhaps the central obsession of human-kind, and indeed of all creatures. We care who our ancestors were, and—probably even more—who our descendants will be. If it were not so, there would be no arranged marriages, no patricians and plebs, no feudalism or apartheid, fewer random beatings in lonely parking lots, and a great deal less genocide. Great swaths of modern law and politics and many thousands of hours of fraught debate would instantly become redundant.

Nowadays, it seems, we understand heredity. Of course, under-standing can never be complete: that is a logical, as well as a prac-tical, impossibility. Any feeling of omniscience that may creep over us from time to time is always an illusion, and a dangerous one at that. Our understanding, nonetheless, already gives us enormous control over the shape, size, color, and even the behavior of our fel-low creatures. Soon we might have crops like metaphorical Christ-mas trees: a basic plant whose species hardly matters (it might be a wheat, it might be a carrot), genetically adorned with whatever extra capabilities and quirks our fancy cares to impose. In a hun-dred years or so (technology takes longer to come on-line than its proponents are wont to suggest!) we might produce livestock like balls of flesh, churning out milk and eggs like termite queens; at least we *could* do this, if we allowed expediency to override sensibility. Most shockingly of all, we could, in the fullness of time, redesign ourselves. We might refashion the human species to a prescription: a height of eight feet for better basketball; an IQ of 400 to talk more freely with computers, win lost and unjust causes in courts of law. Plastic surgery will seem childish indeed when we can restyle ourselves from the genes upward.

Clearly, such technical power—not present-day, but pending in principle—has implications that stretch as far as the imagination can reach. No aspect of economics, politics, philosophy, or religion is untouched by them. What prospect could possibly be more momentous than the redesign of humankind? It is hardly surprising that the notions that have to do with heredity—its theory and the technologies that spring from it—account for much of the content of every modern newspaper. In the past few years in Europe and America, consumers, farmers, politicians, bankers, environmental activists—everyone—has been talking about "GMOs": genetically modified organisms, which, in effect, are the first generation of crops qua metaphorical Christmas trees. The birth of Dolly, the cloned sheep, at Roslin Institute near Edinburgh in 1996 (and the birth the year before, though fewer people noticed, of Megan and Morag) raised the possibility of human cloning. Far more importantly—though most commentators missed this point—the technology of human cloning paves the way for the "designer baby": the human being that is genetically "engineered" to a specification. More broadly, the emotions that spring from matters of heredity continue to ferment and fester: racism and genocide, which surely are as old as humankind (and probably much older), dominate the world's international news.

Yet at the same time, on a more positive note, scientists known as evolutionary psychologists are using what are essentially genetic insights to reexplore the principal theme of the Enlightenment: the true nature of human nature. Already there are encouraging signs that evolutionary psychology can improve on the forays of the eighteenth century, noble as they often were. The point is *not* to be "genetically deterministic," suggesting that human beings are run by their genes, as the critics continue to proclaim. The point is that if we can understand ourselves more fully, then we have a greater chance of devising social structures that are humane and just on the one hand and robust on the other; social structures, that is, that can persist through time and retain their basic humanity and justice through all of life's setbacks and vicissitudes. The criteria of humanity and justice must of course remain the products of human intellect and emotion, as has always been the case: they will not derive directly from greater knowledge of our own biology. But social robustness does depend on such knowledge. There really is no point in devising utopias that require people to behave in nonhuman ways. Many have tried to do this, throughout the twentieth century, and millions died as a result. The possibility of utopia, or something very like it, seems a proper ambition. But it will not be achieved unless we first understand how we really *are*. The vociferous critics of evolutionary psychology should do some homework, to find out what it is really about. If they did, they would surely be ashamed of their own obduracy.

Then, of course, there's the matter of our fellow species: wildlife conservation. I believe this ranks in importance with the fate of humankind itself. The nature of the task for the twenty-first century and beyond is to create a world in which we can thrive *alongside* our fellow creatures. If we succeed at their expense, then this will at least be a partial failure; but of course it is absurd to suggest (as some unfortunately do) that other creatures should survive instead of us, as if human beings should or would commit mass hara-kiri. We and the rest of creation must be catered to, as harmoniously as possible. Wildlife conservation has a huge and necessary emotional content and is indeed driven by emotion, for if people do not *care* about other creatures, then they are doomed. It's unfortunate, though, that some of those who profess to care most deeply also feel that emotion is incompatible with what they see as the cool rationality of science. Yet poetry alone will not save our fellow creatures, either. Wildlife cannot survive without good science. Genetics, the science of heredity, is not the only discipline that's needed, but it is certainly essential. To focus our efforts effectively, we need to know which species are most endangered and in greatest need of immediate help. We cannot begin to make that judgment without knowledge of genes. Sometimes, too, we need to supplement all efforts in the field with specific breeding plans, and these are bound to fail unless guided by genetic theory.

Overall, the present discussion of all these issues, sensible and otherwise, must be welcomed. Nothing can be more important to life on Earth. No human concern is left untouched. In democracies, at least, and most people apparently prefer democracies, we should all talk about and have some input into the ideas and policies that affect our lives. The idea that we should leave everything to "the experts" is an invitation to revert to the worst of the Middle Ages. In the bad old days, the peasantry were expected to take the word of the priest as gospel, while we are now invited to take our lead from scientists and politicians. At least, this is sometimes the case, although some scientists, more sensibly and humbly, are content to take guidance from the societies of which they are a part, and to contribute merely as citizens, like the rest of us. We might indeed argue that it's our *duty* to discuss these issues. The effort is the price we pay for democracy.

However, the issues that now need to be discussed are hugely various: breeding of crops and livestock, cloning, genetic counseling, designer babies, conservation, animal welfare, the nature of human nature. Worse: each of them is, or can be, highly technical. There are entire institutions devoted to each. Specialists in any one field disclaim any worthwhile knowledge of any other. How can the rest of us, who aren't experts at all, hope to keep up? Worst of all: much of the discussion is heavily overlaid with politics of one kind or another. Most perniciously, some biologists seek openly to misrepresent some of the current endeavors, or at least display a sublime lack of understanding; and yet they are believed because they are perceived to have authority. Thus confusion is loaded onto what is already complicated. So what hope is there? Perhaps we should be content to mount discussions in the manner of a party game, to chatter away in the pub for the fun of it, but let "the experts" run the show after all. We could do this, except for the nagging suspicion that the experts aren't as expert as they sometimes make out, and they don't all agree with each other so they can't all be right, and they often seem to say things that leave us feeling uncomfortable. Besides, we don't live in the Middle Ages any more, and although we may feel that priests are important (I certainly do), we should not be content to let them, or scientists masquerading as latter-day priests, tell us what to do. That way of running society is no longer acceptable. But how can we improve on this, if the issues are so disparate and difficult?

Well, I have been looking at aspects of genetics and related issues for about four decades (I am alarmed to discover) and have concluded, after so many summers, that the problems are not as disparate and difficult as all that. It's important only to grasp the underlying principles. Once you see how genes really work—or seem to work, in the light of present knowledge—then all the biology seems to fall into place. The ethical issues require further input, but at least we need not be waylaid by the technicalities. I am not a professional geneticist: I have never plucked the anthers from an antirrhinum, or calculated "gene drops" on a computer, or even tracked some inherited blight through the royal houses of Europe. But I did study genetics and evolutionary theory thoroughly, during my formal education in the 1950s and 1960s, when the world was celebrating some significant centenaries, the new science of molecular biology was undergoing its spectacular birth, and genetic engineering had yet to begin. Over 35 years of work I have had prolonged and significant dealings with medicine, agriculture, and wildlife conservation, and—particularly of late, at the London School of Economics Centre for Philosophy—plenty of involvement with many aspects of genetic and evolutionary theory and the philosophy of science. So I haven't worn a white coat since undergraduate days and am not knee-deep in snails and fruit flies, but I have had the opportunity to take a very long, bird's-eye view of the whole caboodle. It's a matter of swings and merry-go-rounds: what you lose from daily contact with roundworms and oscilloscopes, you gain (I hope) in intellectual mobility. In short, I want in this book to provide an overview of the whole shooting match.

More specifically, I want to show how much the modern world owes to the nineteenth-century Moravian friar Gregor Mendel. As all the world knows, in the 1850s and 1860s Mendel carried out some irreducibly simple experiments with rows of peas, in the monastery garden at Brno. Most people know less about his subsequent experiments—with scarlet runner beans, snapdragons, the hawkweed *Hieracium,* and many others and his final, brilliant flourish with bees—but never mind: peas carry the day. It's often suggested, however, that because Mendel's experiments were simple he was a simple man, driven by simplistic thoughts. Because his results were so precise, too, it has been suggested that he cheated, or—more often—that he was lucky. If he had looked at other plants, or even studied the hereditary pattern of different characters within garden peas, he would not have achieved such tidiness.

But he did not cheat (I can't prove this; but that really wasn't his style, and there are various alternative explanations that seem far more likely). Neither did he rely on luck. He achieved such clarity because he was, in principle, a very modern scientist and knew, as the great British zoologist Sir Peter Medawar observed a hundred years later, that "science is the art of the soluble." While others, including Darwin, pondered the huge complexities and confusions of heredity, Mendel perceived that no one would make any progress until they first identified and worked out the simplest possible cases. He did not stumble upon peas by accident.

He *chose* them from the mass of candidates because he *knew* they would give simple results. He did not study the particular characters he did—the shapes of the seeds and the colors of the pods—by chance. He *knew* before he began that the pattern of inheritance for these particular characters was relatively straightforward. Failure to "breed true" is the bane of all commercial breeders and experimental geneticists: all creatures often give rise to offspring that are strikingly and apparently randomly different from themselves. But Mendel *knew* that his peas would breed true. First, he selected the species for this very quality; and second, he carried out at least two years' preliminary work to produce especially true-breeding lines.

Never, in short, in all of science, have experiments been more beautifully conceived and executed. This is the simplicity not of a simple man but of a genius, who sees the simplicity that lies beneath the surface incoherence; he had the deftness of touch that we associate with Newton, say, or Mozart, or Picasso. Experiments that reveal so much with so little fuss are said to be "elegant." Such elegance—so much derived from so little—has been matched only by Galileo.

Neither did Mendel move on to beans, hawkweeds, and bees out of stupidity. With peas, the simplest possible case, he worked out the ground rules. But he knew, as well as Darwin did, that most cases are more complicated. He wanted to find rules that were universal. I am sure he felt in his bones that the rules he derived from his peas *were* universal, and that with a little tweaking they could explain the odd patterns of inheritance in beans and bees and, indeed, in human beings. But he was one man working more or less on his own, and he ran out of time. In the twentieth century, it took hundreds of scientists several decades to work through the problems Mendel set for himself after his work with peas was done. He also, I think, lost heart—for so few people recognized the greatness of his achievement, and many fail to do so even now.

In truth, the complex patterns of heredity seen in beans and bees and human beings do follow straightforwardly from Mendel's unimprovably simple case. Skin color in human beings is a far more complex character than the shape of the seeds in garden peas, but the theory that explains the latter also fits the former perfectly easily, once we add a few conditional clauses. We see in this some of the essence of the mathematical concept of *chaos:* the notion that very simple rules, once given a small twist, can lead to endlessly various outcomes. Most breeders and biologists were bogged down with the outcomes: Mendel saw the rules beneath. That is genius.

Once we combine Mendel's ideas of heredity with Darwin's concept of evolution "by means of natural selection," we have two thirds of the basic theory of all modern biology. Mendel's and Darwin's ideas together form "the modern synthesis," otherwise known as neo-Darwinism. Add one more set of ideas from the twentieth century—the science of molecular biology, which is the study of what genes are and how they behave—and we have virtually all of the hard core of modern biology. That is not quite true, of course. Ecologists and some psychologists will argue that their disciplines also belong to biology, and that they contain much that cannot simply be thrust into the canon of neo-Darwinism-cum-molecular-biology. But the canon is the hard core nonetheless, the spring from which all may drink.

The English philosopher A. N. Whitehead famously argued that "all moral philosophy is footnotes to Plato": in the same spirit, it could be said that all genetics is footnotes to Mendel. Of all the experiments that he began, only the ones with peas were really completed. But they will suffice. Once we understand what he did, and why he did what he did—his background thought and the details of his irreducibly simple method—*all* the modern debates that seem so confusing start to fall into place.

Chapters 2 and 3 of this book make the general case: they describe Mendel's background, the society in which he worked, and the people he knew; the science of his day, why heredity is so difficult, and how he cut through the knots with his rows of peas. The fourth chapter discusses the twentieth-century contribution: how Mendel's proposed "hereditary factors" were given the sobriquet *genes*, and how the chemistry and modus operandi of the genes were revealed. Chapter 5 describes how scientists worked out what genes actually *are*, and how they operate, leading to the science of "molecular biology."

Then the thesis bifurcates: first into theory, and then into practicalities—the modern biotechnologies. So chapter 6 describes the synthesis, the fusion of Mendel's genetics and Darwin's evolutionary theory, and chapter 7 examines how the notions of neo-Darwinism are now being applied to the Enlightenment issue of human nature through the discipline of evolutionary psychology. Chapter 8 then discusses the practical issue: how modern genetic theory is applied to the "improvement" of crops and livestock for economic purposes, and, in sharp contrast, how it can be pressed into the service of conservation.

Chapter 9 shows how the two arms of the discussion, the deep theory and the practice, are being brought together again to make the most fundamental changes to human life that are conceivable. It is rapidly becoming possible to redesign human beings: to create the "designer baby."

We have already reached the point, indeed, where we might reasonably argue that all the conceivable ambitions of biotechnology should be considered possible, provided only that they do not break what Medawar called the bedrock laws of physics. There is nowhere for biotechnology to hide, in short. Scientists can no longer dismiss public misgivings with the peremptory comment that this bogeyman or that need not be considered because it is "not possible" or mere "fantasy." Such dismissals just will not do. *Anything* that a physicist would allow, the biologists of the future might do. When anything is possible, we have to ask as a matter of urgency: So what is *right*? That is the subject of the last chapter.

Mendel has much to teach us here, too. Much less is known of his life than we should like. Most of his personal notes were destroyed. The literature of Mendel ought to be as rich as that of Darwin, and it simply is not. So we do not know in detail much of what he thought or did, day to day. The little that is known, though, suggests that he was courteous, humble, generous, and certainly conscientious. These are qualities our descendants will need in abundance if they are to come to terms with their own extraordinary power.

TWO

The Peasant and the Scientist

Modern biology began in the nineteenth century with two great contrasting traditions. One was English and emphatically rational—in some ways more obviously modern—yet out of it came the great metaphysical overview of Charles Darwin: the notion of evolution "by means of natural selection." The other was German. This tradition was more "Romantic" in the literal sense, pervaded as it was by the quasi-mystical insights of *Naturphilosophie* (which so hugely influenced the Romantic movement, as manifest most obviously in Coleridge), yet it gave us the more down-to-earth arm of modern biology. The German scientists showed how living systems actually *work*. Notably, in the early nineteenth century, they revealed how all large creatures, such as animals and plants, are compounded of many tiny cells, and how those cells divide and develop into embryos and hence into new, free-living individuals. Later in the century they invented biochemistry.

At the pinnacle of the Germanic tradition stands Gregor Mendel. By no means unaided, but out on his own nonetheless, he effectively created genetics, the science of heredity. The only experiments that he truly took to completion were with garden peas, which he grew in the monastery garden at Brno in Moravia, now in the Czech Republic, not far from Prague. All that has happened in the name of genetics in the twentieth century, and everything that is happening now, falls easily into place with a basic understanding of Mendel's ideas.

The textbooks generally assert that Mendel was Austrian, presumably because Moravia, where he was born and lived most of his life, then formed part of the Austro-Hungarian Empire. Ethnically, however, he was German. He and most of his family spoke German; he learned Czech, the local language, fairly late in life and not particularly well. Genetics is, of course, a branch of biology, but Mendel's first love was physics. His scientific friends and mentors included such greats as the physicist Christian Johann Doppler, who if he had lived a century later would certainly have won a Nobel Prize. The Doppler effect explains why the pitch of a police car's siren drops as the car whizzes past, which might not seem particularly exciting; but this insight, applied to light rather than sound, has become one of the key devices of modern cosmology, enabling astronomers to calculate the speed at which distant stars are retreating from us, and hence to judge their distance. Such were the social and bureaucratic arrangements of mid-nineteenth-century Europe, that Doppler at one point found himself interviewing Mendel for a job as a small-town schoolmaster.

Mendel did indeed teach science (though he never actually managed to pass the teaching exams), but he lived all his adult life in the monastery at Brno. Since his order was Augustinian, he was technically a friar, although he is usually described simply as a monk. It is not clear that he was particularly devout: the sparse personal notes that have been left suggest that he entered the monastery largely for economic reasons, and took holy orders because that is what monks did. But he was certainly conscientious, and for the last 17 years of his life he was abbot of Brno, fighting the monastery's corner against the Catholic bureaucracy.

His life, though, and to some extent its aftermath, remains largely mysterious. After he died in 1884, just 61 years old, a meddlesome monk burned most of his notes, including those of many of his experiments. Why? The gesture may have been well-intentioned, but in retrospect it seems the most extraordinary act of vandalism. In 1865, Mendel twice presented the results of his seminal experiments with peas to audiences that must have included some of Europe's leading biologists, and in 1866 he published his findings in a paper that is a model of thoroughness and clarity.

Yet his work was apparently ignored. He entered a long, fruitless, and entirely dispiriting correspondence with a Swiss botanist, Carl Naegeli (famous in his time, but now more or less forgotten), and carried out many more experiments with different plants. Finally, toward the end of his life, he carried out brilliant experiments with bees, in which he kept swarms of different varieties under perfect control. He arranged marriages between the drones of one colony and the queens of another and then studied the resulting hybrids. Even nowadays, such experiments would require great skill. Given the technical limitations of the day, and the fact that at the time Mendel was a full-time abbot, they are breathtaking. But still no one took any serious notice. His research disappeared from view for several decades, until his one published paper, of 1866, was rediscovered around 1900 by three biologists, working independently: the German Carl Correns, the Austrian Erich Tschermak, and the Dutchman Hugo De Vries. Then the science of heredity that Mendel had so ably founded—soon to be named genetics—could properly begin. But when the English biologist William Bateson visited Brno in 1904 to find out more about the strange young monk who had done so much and then been lost to view, no one could throw serious light on his work. They remembered the donation he had made to the fire station at Hyncice (the village of his birth), and that he had been amiable, courteous, hardworking, and well loved, if a little tetchy at times (particularly with the Catholic bureaucracy). But peas? Hmm. Yes—I think you'd better ask, um, somebody else. Bateson returned to England, nonplussed.

Bit by painful bit, however, the bones of Mendel's life, work, and education have been unearthed. His nephews, Alois and Ferdinand Schindler, provided much of the background. Their mother, Mendel's sister, had given Mendel part of her dowry to help with his education, and he in turn supported her sons through theirs. H. Iltis (1882–1952), a natural scientist from Brno, filled in many of the remaining gaps and published a biography in 1924. My account leans heavily on the excellent biography by Vitezslav Orel: *Gregor Mendel: The First Geneticist.* Thanks to these scholars and others, much, though by no means all, of the mystery has been resolved. At least it's clear now who Mendel actually was and what he was doing working in a monastery, for monasteries, in today's world, are not usually associated with seminal science. It's clear too that his experiments with peas were beautifully designed, wonderfully modern in concept, and rooted in a deep understanding of the problems—problems that now seem difficult to envisage largely because Mendel solved them. Indeed, we can hardly doubt that Mendel should be ranked among the pantheon: the elite that includes Galileo and Newton among physicists, and Darwin (it is hard to think of another quite so significant) among biologists.

It is still not obvious, however, why Mendel's many excellent friends and acquaintances did not appreciate him more than they did, and why they allowed his work to disappear, to be rescued only by the skin of its teeth. Neither is it clear why Mendel's ideas apparently caused so much angst after they were rediscovered; why some, including Bateson, found it hard to reconcile Mendel's genetics with the newly emerging knowledge of chromosomes, the structures that physically contain the genes, or with the "biometric" tradition that had grown up in the late nineteenth century with Darwin's cousin, Francis Galton; or why, until the 1920s and beyond, it seemed so hard to reconcile Mendel's ideas with Darwin's. By the 1940s Darwin's and Mendel's ideas were finally brought together to produce "the modern synthesis," otherwise known as neo-Darwinism, and this was one of the great intellectual triumphs of the twentieth century. But it seemed devilish hard work to get to that point.

SO WHO WAS MENDEL?

Mendel's life story is often related in the traditional manner of mythology: born in obscurity; self-taught; playing with plants in a monastery garden for no apparent reason, except to while away the long, quiet summers; hitting on experiments that changed the world by lucky accident. In almost all respects the myth is wrong, as most myths are. In truth, he was born to intelligent and very able people, and into a society that was obsessed with the problems of agriculture, particularly crop improvement; it had to be, because its livelihood depended on it. He was well educated, and extremely appropriately, primarily as a physicist and mathematician; and he brought the precision of those austere disciplines to the problems of heredity, as others had recommended before but rarely attempted. Indeed, it's almost as if he was destined to do the work he did: exactly the right kind of thinker in the right place at the right time, surrounded by fine scientists but far ahead of all of them. His family made sacrifices on his behalf, and a succession of fine teachers, clerics, and scientists helped him on his way: Schreiber, Franz, Napp, Keller, Diebl, Klacel, Bratranek, Baumgartner, Doppler, Ettinghausen, Unger. He was also acquainted—at least at one step removed—with the greats of nineteenth-century European biology, including Schleiden, Purkinje, Virchow, and even Bunsen and Liebig. It is strange that his career as a scientist seemed to peter out so limply in the correspondence with Naegeli. Why, when he knew so many fine people, did he spend so much time on one who clearly did not appreciate his work? Truly, Mendel's life and work need no mythologizing; but there is mystery in their denouement nonetheless.

Mendel was born in 1822—on July 22, though the baptismal register says July 20—in Heinzendorf (now Hyncice), a little village of 72 households. He was christened Johann and took the name Gregor later, when he became a monk. The Mendels, evidently from Germany, had been in the area since the mid-1500s. In the Austro-Hungarian Empire of the early nineteenth century, the prevailing politics were feudal, for feudalism officially came to an end only in the revolutionary year of 1848. Mendel's family was of the kind sometimes classified as rich peasant: people of no obvious political clout, but essential to the economy and not to be ignored. Mendel's father, Anton, was a farmer who owned the family's house and improved it, replacing the original wood with brick. Inevitably, as a good Moravian, he grew fruit trees in the garden and kept bees. But he also owed allegiance to the feudal lord in whose fields and forests he was obliged to work three days a week, with his two horses alongside him. Johann's mother, Rosine (née Schwirtlich), was the daughter of market gardeners. Religion was a powerful force, too. Moravia, like most of northern Europe, endured alternate bouts of Catholics and Protestants. Through Mendel's own life the Catholics prevailed.

The feudal regime was oppressive but not generally brutal. The ruling class at its best acknowledged the ancient principle of noblesse oblige and, much of the time, sought to look after the peasantry: benign patronage. The Catholic Church the world over often seems to lie like the wettest of wet blankets over all intellectual adventure: the defender of dogma, the natural enemy of science. So it did in nineteenth-century Moravia, just as in seventeenth-century Italy, when it put a stop to Galileo. But if only the coin had flipped the other way, the Catholic Church *might* have emerged as the champion and bastion of science, and the course of Western philosophy and science would then have run quite differently. Galileo, after all, was a devout Catholic. The Jesuit intellectuals were entranced by his cosmology; it was the perceived threat to papal authority that they could not endure. The monasteries in general, in the centuries before Galileo, gave birth to modern science. Though they had suffered mightily at the end of the eighteenth century, the monasteries of central Europe during the nineteenth century were still perceived as hives of intellectual activity, and science (embracing agriculture) was high among their obsessions. The monasteries might be seen as the precursors of modern research stations, though all their research was carried out within the context of Catholic devotion. To a large extent, the particular intellectual pursuit of each monastery was determined by the predilections of its particular abbot. Even so, in nineteenth-century Moravia, as in seventeenth-century Italy, any true spark of intellectual adventurousness was liable to be stamped out. There were plenty of avant-garde thinkers and teachers at the monastery of St. Thomas's at Brno, where Mendel spent his professional life, and a striking proportion of them lost their jobs.

On the other hand, Moravia as a whole was emerging as a modern economy rooted, as usual in the early nineteenth century, in agriculture. There were two prevailing obsessions, which is hardly too strong a word, for the whole economy depended on them. One was sheep (mainly for wool) and the other was fruit, including vines for wine but also apples and other fruits as significant sources of nutrition. Agriculture was already competitive internationally—the rise of Britain's empire threatened Moravia's ascendancy in wool—and absolutely everyone knew that success would lie with those who could produce the *best* sheep and the *best* crops. In short, the national obsession with agriculture was underpinned by a national obsession with the breeding of livestock and crops. Brno was the capital of Moravia, and at that time it was a rising industrial town—the center of the Hapsburgs' textile industry in a region where wool was at the core of the economy. In the early nineteenth century it established its own Agricultural Society, as well as two more specialist groups: the Sheep Breeders' Association and the Pomological (apple-growing) and Oenological (wine) Association. These were serious centers of intellectual activity. In 1849 the Agricultural Society created a Natural Science Section, which in turn, in 1860, achieved independence as the Natural Science Society. Mendel and most of his scientific friends were members, and the society's 24 honorary members included many of the greats, such as Bunsen, Unger, Virchow, and Purkinje.

Thus, alongside feudalism and religious bureaucracy—and to some extent the product of them—Moravia had a significant intellectual and educational tradition, very much with a practical and scientific edge, as befitted that agricultural and industrial society. Throughout its history can be seen the same kind of fervor for self-improvement, and the enthusiasm of a few outstanding reformers to meet that desire, that ran through Britain and was typified by the workers' educational schemes of the nineteenth century. Mendel, both the young Johann and the mature Gregor, benefited enormously from the zeal of earlier Moravian reformers.

In the seventeenth century the great Protestant liberal John Amos Comenius insisted that education should be general, and should include natural history: he was, after all, a follower of Francis Bacon. Inevitably, it seems, Comenius was forced into exile in 1620 when the Catholics returned; but he made his mark nonetheless, and Mendel, two centuries later, was a beneficiary. In the era just before Mendel was born, at the end of the eighteenth century, the Countess Walpurga Truchsess-Zeil, supporter of the Freemasons and of the Enlightenment, founded a private institution of learning at Kunin, where she paid bed and board for local children of talent, both boys and girls. Science featured conspicuously in the education. One of the teachers was the outstanding reformer Christian Carl Andre, who also had a great influence on the region as a whole; another was J. Schreiber, who went on to shape the life of the young Mendel in particular.

Schreiber (1769–1850) was an innovator—gardener, teacher, social reformer, and priest. At Kunin he established a nursery for fruit trees—for what else would a Moravian do?—with varieties imported from France. He and the countess sought to distribute the best strains among the villagers, but they knew that if they simply gave them away, the villagers would not value them. So Schreiber and the countess put the word about that anyone caught stealing the new, world-beating, imported seedlings would be severely punished. They posted guards with strict instructions to make the loudest possible noise but never actually arrest anybody. Within three days all the seedlings had been stolen. In our own times, there have been comparable strategies to distribute contraceptives in the Third World. Official channels are simply not trusted. The black market is the vernacular route. We may note in passing the benign face of feudalism: the positive desire to help the people at large. Modern freedom is of course to be preferred, but with it has come the hard-faced market, where everything with an edge is patented and nothing can ever be given away.

But the dank fog of bureaucracy soon closed in. The Austro-Hungarian authorities liked the school at Kunin well enough at first but then became frightened by the Revolution in France. They investigated the teaching at Kunin, found it too "liberal," took particular objection to the "alien notions" of science, and held Schreiber responsible for what they called the scandal. Schreiber was forced to leave in 1802, and Kunin closed in 1814. The whole political scene, each lurch toward liberality and modernity brutally and predictably checked after just a few years by some conservative and nervous regime, is all too reminiscent of modern times. We have seen it in the old USSR, in present-day China, and—lest we are tempted to be chauvinistic—in modern Kansas, where the teaching of evolutionary theory has lately been suppressed. Kunin's misfortune, however, was to benefit Mendel. The sacked Schreiber became the priest of the parish that included Mendel's native village of Hyncice.

Anton Mendel, Johann's father, was born in the first year of the French Revolution, 1789, and took part in the last stages of the Napoleonic Wars. He married Rosine Schwirtlich, the gardener's daughter, in 1818. She was five years his junior. The Mendels' first child, Veronica, was born in 1820; then came Johann—their only son—in 1822; then Theresa, in 1829. Two other children died soon after birth.

Anton clearly flourished for a time. Like all the people of Hyncice, the Mendels wanted their children to "get on," and to "get on" meant to be educated. There were even pedagogues in the family: Rosine's uncle had taught at the village school.

And Schreiber was already on the scene. As he had done at Kunin, he established a nursery of fruit trees in the presbytery garden at Vrazne. In 1816 he helped found Brno's Pomological Association; by then, too, he was a corresponding member of the Brno Agricultural Society. Schreiber taught young Johann the basics of fruit tree improvement. Among other things, the key technique of artificial pollination had been in wide use in Europe since about 1820. In 1833, when Mendel was 11, Schreiber and the schoolmaster Thomas Makitta commended him to the school in Lipnik (Leipnik), where gifted pupils were primed for the *Gymnasium*—the secondary school—at Opava (Troppau). One year later the 12-year-old Johann duly graduated to the *Gymnasium*.

SCHOOLDAYS

But Johann's success was hard on his parents. Opava was more than 20 miles away, and he had to live in, though his parents could only afford to pay half board. Furthermore, they had lost the natural heir to their farm. Then came an unkind twist. In 1838, when Johann was in his fourth year at Opava, Anton was seriously injured in the forest while working for the landowner. Johann was then only 16, but already, as he later recorded in his curriculum vitae, he was "in the sad position of having to cope for himself entirely." (Quaintly, Mendel always wrote about himself in the third person.) The stress evidently wore him down, for he fell seriously ill and had to spend much of the following year, 1839, at home. He was bright, though, and he still went up to the next class. He also took the first formal steps to become a schoolteacher—his prime ambition—and attended a course at the teachers' seminary at Opava. Again he did well and was able to earn what he later called a scant living, teaching his less able schoolfellows. But life was hard. As he also recorded later: "His sorrowful youth taught him early the serious aspects of life, and it also taught him to work."

Johann remained at Opava for six years, and apart from his periods of illness, he returned home only during the holidays. Taken all in all, though, Opava was not a bad place for a young biologist. It had a museum of natural history (established in 1814 at the suggestion of Christian Andre) and a library that included books on meteorology, which became one of Mendel's lifelong passions.

Still, though, Johann was not ready for university. To qualify, students had to complete two more years at what the British would call a sixth-form college: either a lyceum, which were attached to universities, or an independent philosophical institution. The nearest such institution was at Olomouc (Olmütz), and so, in 1840, Mendel offered his services as a private tutor in Olomouc so that he could continue his studies. But he failed to find work and again became sick. So followed yet another year at home with his parents. Again, though, life was not entirely bleak. At Olomouc he met P. Kriztovsky, who achieved fame—at least locally—as a composer. At that time Kriztovsky had just left the Augustinian monastery and was trying to find work as a teacher. Throughout his life, Mendel made good friends outside science as well as within. (Much later, as abbot at Brno, he employed the great Leoš Janáček as organist and choirmaster.)

OLOMOUC

Johann, now aged 19, finally joined the Philosophy Institute at Olomouc in 1841. He was pulled two ways, though, because his injured father, no longer able to work, asked Johann one last time to return home and take over the farm. Johann stayed where he was, however, though he must have been tempted to leave: he was in a strange city with no old friends and little money, and he spoke the local Czech language very poorly. Indeed, he suffered yet another mental crisis in his first term and again went home to recover, but by then he had already passed Latin and mathematics with top grades. So he returned to Olomouc to complete the course. At this point, he at least acquired a safety net. His elder sister, Veronica, married, and her new husband bought the Mendel family farm. The sale of contract, dated August 7, 1841, includes a clause that was to supply Johann with a small annual allowance if he entered the priesthood, while rooms were to be reserved for him at the farm if he failed. His younger sister, Theresa, then offered Johann part of her dowry to help him out, and Mendel later paid her back by supporting her three sons in their studies.

At Olomouc, Mendel was examined in religious studies, philosophy, ethics, mathematics, physics, and pedagogics. He did *not* take natural history and agriculture but specialized instead in mathematics and physics: seven hours a week of math in his first year, out of a total of 20 hours, and eight hours a week of physics in his second year. He finished the course in 1843 and could then have gone to Olomouc's sixteenth-century university. But—presumably through lack of funds—he did not. Instead, his physics teacher at Olomouc, F. Franz, recommended Mendel to Abbot F. C. Napp at the monastery of Brno. Thirteen candidates applied for the novitiate, and Mendel was one of four to be short-listed. Franz was warm in his praise: "During the two-year course in philosophy [he] almost invariably had the most exceptional reports." He was, said Franz, "a young man of very solid character—in my own branch almost the best." By "my own branch," of course, Franz meant physics, which at that time seemed Mendel's destiny.

LIFE WITH THE MONKS

Napp accepted Mendel without interview on September 7, 1843, and his parents gave their written permission on September 19. He began his novitiate on October 9, 1843, now aged 21 years and nearly three months. It was at this point that he took the name Gregor. It pleases me to think of him in the great tradition of biologist-clerics, such as England's Gilbert White, whose *Natural History and Antiquities of Selbourne* of 1788 was such an inspiration to later biologists, including Darwin. In truth, though, Mendel leaves no evidence of serious religious commitment. Mendel recorded in his curriculum vitae of 1850 that he entered the monastery because he was tired of overwork. "It was impossible for him to endure such exertion any further. Therefore, having finished his philosophical studies, he felt himself compelled to enter a station in life that would free him from the bitter struggle for existence. His circumstances decided his vocational choice. He requested and received in the year 1843 admission to the Augustinian monastery of St Thomas in Brno."

Yet life at Brno was in many ways ideal. Mendel was a natural scholar and, as he tells us in his curriculum vitae, ". . . so long as he fulfilled his clerical duties, he was free to devote himself to private study," while he had, he said, the security that was ". . . so beneficial to any kind of study." Many a modern scientist would love to claim as much. As at Opava, he had specimens on hand, and in his spare time ". . . he occupied himself with the small botanical and mineralogical collection which was placed at his disposal in the monastery." As he gave more and more attention to natural science, he developed "a special liking, which deepened, the more he had the opportunity to become familiar with it."

As for Napp, he had become abbot in 1821, when he was only 32, and he remained in the post until his death in 1867, when Mendel took over. As Vitezslav Orel records, Napp was a passionate student of philosophy and theology, who had wanted to be a university professor. He was also—inevitably, it seems, in nineteenth-century Moravia—a plant breeder. Before he came to Brno he had organized a nursery garden in the monastery's farm at Sardice, 50 miles away, where he produced new strains of apples; in the 1830s he created an experimental garden at the Brno monastery that was described as "a jewel among nurseries." Indeed, Napp's job description required him to "promote pomology and wine-growing through experiments, observation, and instruction, enrich science, and propagate useful findings." Mendel himself, as a novice, was placed in the care of A. Keller (1783–1853), yet another member of Brno's Agricultural Society and an active member of the Pomological Association. Keller sought primarily to improve vines and fruit trees, and had published articles on melons.

The monastery itself had been founded in 1350 next to the church of St. Thomas, and for time it flourished. It was rebuilt in the seventeenth century with a large library and accommodation for 42 monks. But then, in 1782, more than half the monasteries of Europe were abolished, while those that remained had to serve the state as well as the church. The monks were now obliged to work in parishes and hospitals and to teach religion in schools. Brno survived, but in 1783 the monks had to move from their fine building in the center of town to a former convent on the outskirts, recently vacated by Cistercian nuns. The formerly affluent monastery was now in debt. The monks' new home had been rebuilt in the fourteenth century and was in poor condition. It was still run-down in 1807 when orders came from Emperor Franz I (1792–1835) for the monks to teach math and biblical studies in the newly established Philosophy Institute and Brno's theological college.

So the monasteries actively had to recruit young men to teach: men just like Mendel. P. F. Nedele (1778–1827) was the first teacher of biblical studies at Brno, but the bishop considered him too liberal because of his involvement in the Czech national revival movement, and he was fired in 1821, whereupon he became the prior and librarian. Napp then took over as abbot.

Napp too was constantly at odds with the Catholic authorities, who wanted him to cut down on teaching and focus on monastic discipline, but he had influence as a member of both the Lords Diet and the Brno Agricultural Society and was able to resist. He welcomed the imperial order for monks to teach and insisted that they should teach science. He also had plans to rationalize agriculture, and to this end he enlisted the advice of F. Diebl, whom he helped to become professor of agricultural science at Brno University. Diebl advised Napp on crop rotation, the cultivation of fodder crops, and the breeding of sheep. The monastery, after all, obtained much of its income from the farm at Sardice, with most of the cash coming from wool. Napp carried out field trials, and in 1826 he and Diebl analyzed the reasons for a particularly bad invasion of pests, which they showed resulted from the warm autumn. In 1827, Napp was elected president of the Pomological Association, and he was also involved with the Sheep Breeders' Association. Of specific relevance to Mendel, Napp pointed out in 1836 that the problems of heredity could be solved *only* through experiments. All in all, Napp was a deep thinker and a fine scientist. Mendel may have entered the monastery at Brno primarily to escape privation, but the place, and the people in charge of it, could hardly have been better suited to his development.

Mendel acquired the plot in which he grew his peas—perhaps the most significant allotment in the history of humankind—by a route that also reflects the essential seriousness, and the special concern for science and agriculture, in nineteenth-century Moravia. The whole tradition was largely established by Christian Andre, but Moravia's first notable plant scientist was A. Thaler (1796–1843), yet another Augustinian, who taught mathematics at the Philosophy Institute. In 1830, Napp gave Thaler permission to establish an experimental garden in the monastery, and he created a substantial one of 35 yards by 5 yards, under the refectory windows. There he grew rare Moravian plants and founded a herbarium.

Matous Klacel (1808–82) took over the experimental garden from Thaler in 1843, the year that Mendel joined the monastery. Klacel was primarily a philosopher who studied natural science. He was a follower of Hegel and embraced the German system of *Naturphilosphie,* which demanded that the natural philosopher should become emotionally involved in his subject. Indeed, said Klacel, "deliberate university is the aim of all love and science." Such a view these days might be subsumed in the general notion of "holism," though the currents of *Naturphilosophie* ran very deep, essentially mystical. Klacel too was a member of the Agricultural Society, and between 1841 and 1843 he published articles on the need for progress in natural and social sciences. He also espoused the cause of Czech nationalism, which aroused the ire of the authorities, and in 1842 and 1843 he was accused of spreading pantheistic notions, and then the ideas of the philosopher Hegel, in opposition to the Catholic faith. Napp defended him as best he could, but in 1847, Klacel, in the eyes of the Catholic authorities, finally went too far. He wrote a three-part essay, "The Philosophy of Rational Good," the official censors in Vienna declared that the third part contained "harmful sentences," and (how drearily repetitive this becomes!) he was dismissed from his post as philosophy teacher.

The following year, 1848, brought revolution throughout Europe. Klacel helped organize the pan-Slavonic congress in Prague, and it was then that he asked Mendel—the two had long been firm friends—to take over the experimental garden. The congress was broken up by force, and Klacel returned to the monastery to become the librarian (there is a thesis to be written about mavericks of all kinds who take refuge in librarianship). Nonetheless, on March 13, 1848, the absolutist government of Metternich was forced to resign, and a new—capitalist—order arose. Emperor Ferdinand I promised to abolish feudal labor, to lift censorship, and to summon a legislative assembly. On March 17 there was a mass meeting in Brno; a revolutionary poem set to music by Krizhovsky was sung, as they marched, by the newly formed revolutionary guard. Diebl marched among the students at the head of the procession. Napp was a supporter and also entertained revolutionary students from Vienna. Klacel gave a sermon, demanding the end of feudal labor, though the city's captains of industry objected and, in turn, demonstrated against the monastery. The government in Vienna finally abolished formal labor on September 7, officially ending feudalism. All in all, 1848 was a momentous year. It changed the face of Europe; and Mendel acquired the monastery garden in which he changed the world.

Eighteen forty-eight was also the year in which, on August 6, Mendel received holy orders. He had already been given an ecclesiastical post, as curate to a parish attached to the monastery. Nonetheless, he and five of his fellow monks signed a petition for the reform of the monasteries, pointing out that monks were effectively deprived of civil rights, like patients in a mental hospital. The petition, inevitably, was written by the fiery Klacel, then aged 40. Klacel described the monasteries of the empire as "almshouses for poor and short-sighted youths" (did he have Mendel in mind?) who came to know "enforced isolation" and lived in "the nadir of degradation." The petitioners demanded "free, united, and indivisible citizenship." The petition found its way to Vienna and there, apparently, was lost, for about a hundred years. It did not turn up again until 1955.

Mendel's curacy in the monastery parish began on July 20, 1848, a fortnight or so before his ordination. His parishioners included the sick at a nearby hospital. He was, as all who knew him agree, a kindly man and well organized, but the priesthood did not suit him. He wanted freedom to teach and to study science. Once more—as so often in times of stress—he fell ill; in January and February 1849 he was confined to his bed for 34 days. The ever accommodating Napp rescued him from his misery. In the summer of 1849, he recommended Mendel for a post as substitute teacher in math and classics for the seventh-grade pupils at Znojmo (Znaim) in southern Moravia.

So Mendel became a teacher on October 7, 1849. He had to borrow money to survive at first, but he had a salary nonetheless, of 360 guilders a year. He did not, however, have a professional teaching qualification, and so, as a new law demanded, he had to take an exam at a university to prove his competence. On April 16, 1850, the headmaster at Znojmo applied for Mendel to take the exam, to teach natural history to all grades, and physics to the lower grades. The headmaster thought highly of him for he had, he said, a "vivid and lucid method of teaching."

But the exam was difficult; Orel tells us that it would have needed several years of full-time preparation, and although Mendel worked hard for it, he was already teaching 20 hours a week. One of his essays, on meteorology, went down well. It was on the properties of air and the origin of wind, and discussed the possibility of forecasting weather. On this, he was at one with Robert Fitzroy, Darwin's captain and companion on the voyage of the *Beagle;* Fitzroy devoted his last years to weather forecasting. The examiner, Professor Baumgartner, who was then minister of trade, liked what Mendel wrote.

But his second essay, on geology, fell foul of its examiner, R. Kner, yet the essay survives and seems very fine. Later scholars suggest that Kner was biased against Mendel because he was a monk. But perhaps he was simply too revolutionary. Following Charles Lyell (1797–1875)—the geologist who so inspired Darwin—Mendel wrote that "the history of creation is not yet finished": evolutionary change continues to unfold. No modern scientist would deny this, but it offended the mid-nineteenth-century conservative notion that God's work was completed with His creation of man. In the event, Kner concluded that Mendel, one of the great scientists of all time, was not of a standard to teach natural history to 16-year-old pupils in the upper grades of a *Gymnasium.*

So with one success behind him, and one failure, Mendel proceeded to further examination in Vienna in August 1850. His examiners in physics again included Baumgartner, and the great Christian Doppler. Mendel's essay on the making of a permanent magnet impressed them both. Again, though, he came up against the recalcitrant Kner, who took exception to the way he described the classification of mammals. Kner had written a book on this, unknown to Mendel, who offered quotes from a different book. So he failed again. This time one of the world's greatest biologists was declared unfit to teach natural history even to the lower grades in the *Gymnasium*. This failure, Mendel later wrote, was a "stunning disappointment."

So he went back to Brno to become a substitute teacher at Brno's brand-new Technical School, which had been established in 1850. He did not need specialist qualification for this; although without the formal certificates, he did have to make do on half pay. At the school he was commended by the headmaster and well liked by the pupils and his fellow teachers. He was also admitted as an extraordinary member of the newly established Natural Science Section of Brno's Agricultural Society. The technical school went on to become an Institute of Technology, in 1867, and then, in 1873, the Technical University. With feudalism now officially buried, there was a bullish demand for more professionalism in agriculture and industry.

At this point, Mendel experienced yet another helping hand. Professor Baumgartner, whom he so impressed in Vienna, and who was presumably irritated by Kner, wrote to Napp to say that Mendel should be allowed to study full-time at the University of Vienna. Napp agreed—perhaps he had already thought of this himself—and on October 3, 1852, he wrote to Baumgartner to say that he had decided to send Mendel for "higher scientific training" and to this end would "regret no expense." In addition—no doubt covering his tracks, but perhaps in some exasperation—Napp wrote to the bishop of Brno to say that Mendel would in any case have made a lousy parish priest. The mayor of Znojmo asked Mendel to return, but Napp replied that he had "other plans." So we should be grateful to Kner: but for him, and a later examiner called Fenzl, Mendel would surely have whiled away his days as a full-time small-town schoolmaster and never been heard of again. Kner himself seems to have passed into obscurity. His cussedness, however, liberated Mendel to become one of the world's most significant experimental scientists.

VIENNA

So in 1852, at age 29, Mendel enrolled for full-time study at the University of Vienna; there he came under the sway of physicists and biologists who, by any standards, must be considered outstanding. To the peasant upbringing in the wiles and importance of gardening and breeding, he now added the techniques and insights of modern science. For although the uprising of 1848 did not bring all the changes the reformers wanted, it did transform university education. Vienna had acquired a new faculty of philosophy—which included science—and Napp sent Mendel to study physics under Christian Doppler, who became head of the new physics institute in January 1850. The physics building was finished in 1851 and was designed to accommodate twelve students, though Mendel was in fact the thirteenth. The course was to last three and a half terms, but Mendel signed up for another, possibly because Doppler died in 1853 (only 50 years old) just as Mendel was beginning his second academic year. A. Ettinghausen (1796–1878) took Doppler's place. Mendel still considered himself to be, above all, a physics teacher, and physics was widely considered to provide very good grounding both for teachers and for industry. Later, however, he also signed up for math, chemistry, zoology, botany, the physiology of plants, and paleontology. He canceled zoology, however, because the lessons clashed with physics. The students worked very hard. Math lectures took place on Saturdays—with physiology practicals on Sundays! For light relief, Mendel bought lottery tickets. He was a down-to-earth fellow with an eye for statistics—the fancy word for "odds."

Physics and math, it may appear, have little to do with plant breeding and the science of genetics. Yet they were crucial: Mendel succeeded in making sense of heredity because he brought the precision and purity of physics to the subject. Naturalists like Darwin, who seem more obviously suited to the task, are too keen to pursue the myriad variations of nature, but the task is to find the underlying order. Both Doppler and Ettinghausen were wonderfully appropriate teachers. Doppler above all emphasized *economy* of experiment, that researchers should seek the most apt (we would say "elegant") experimental approaches to their problems. Doppler had also published a book on math, including a discussion of probability. Ettinghausen had taught at Vienna since 1821, and he too had published books on math, including a textbook, *Combinatorial Analysis,* in 1826. Thus Mendel learned—as Darwin, for example, never did—the absolute importance of statistics when seeking the patterns that lie within variety.

At the same time, though, and crucially, Mendel was introduced to serious botany at Vienna. It acquired the status of a specialty during Mendel's time. Experimental plant anatomy and botany were taught by yet another renowned scientist, Franz Unger, who lived from 1800 to 1870 and taught at Vienna from 1848 to 1866. He was also an honorary member of the Brno Natural Science Society. Unger, like Doppler, emphasized the need for sharp experiments, and he was keen on the general approach of *Naturphilosophie,* but also on cytology, the study of cells. Indeed, he was influenced in this by J. M. Schleiden (1804–1881), who introduced the idea of the plant cell and wrote *Principles of Scientific Botany* in the years 1848 to 1850. Schleiden, in turn, was a pupil of Professor F. Fries (1773–1843), who had emphasized the mathematical philosophy of nature. Schleiden wrote in his *Botany:* "a complete theoretical explanation [of natural phenomena] is possible only on the basis of mathematics, and only in so far as mathematical treatment is feasible." So even the botany that Mendel learned had a mathematical pedigree: from Fries, through Schleiden, to Unger. Schleiden also had views on plant hybrids (crossing plants of different kinds), and Mendel's pea experiments were essentially an exercise in hybridization. As Schleiden wrote in 1848, ". . . embryo formation [is] determined from both sides [parents] and in a way represents the mean." All along the line, Mendel seem to have been primed for his task.

Unger was a fine botanist. Nature is much more confusing than we might choose to think it is, and it is much harder than it seems to find out anything for certain. For example, while it is obvious that plants of any given kind vary, is the variation due to innate differences or simply to the environment? Is this nature or nurture? The only way to tell is by growing plants that are genetically identical, to put the matter in modern parlance, in different soils. The matter is easily resolved today: now that we understand the mechanisms of inheritance, and can minutely adjust soil chemistry and other conditions. It was much harder in the mid–nineteenth century. It was certainly known then that plants of a given type vary according to the conditions in which they are grown. But Unger was able to show that they also vary for reasons *other* than soil type. As we would say today, he was able to reveal the variation that is due to genes. Mendel himself transplanted plants to pin down the source of variation. It is, of course, impossible to make sense of the rules of inheritance if the variations brought about by environment are not taken into account.

Unger was also among the coterie of biologists who were considering evolution well before Charles Darwin wrote *Origin of Species*. In 1851 he wrote a book in which he described extinct plants and spoke of life developing from "lower" forms to "higher," until man appeared. He rejected the idea—as Darwin also did, of course—that species are constant. "Who can deny," he said, "that new combinations arise out of this permutation of vegetation . . . which emancipate themselves from the preceding characteristics of the species and appear as a new species?" Unger studied hybridization—the theme closest to Mendel's own studies—and pointed out that artificial pollination was useful in creating new varieties. He wrote about all this in 1860 and mentioned species that Mendel himself made use of, including various fruit trees and fuchsias. Hugo de Vries (1848–1935), one of the "discoverers" of Mendel's work, quoted Unger in his own thesis on Mendel: "It is the task of physiology to reduce the phenomenon of life to known physical and chemical laws."

However, in the mid-nineteenth-century Germanic tradition (and in the tradition of *Naturphilosophie*), Unger tended to conflate the changes seen from generation to generation in lineages of creatures—what we now call evolution—with the kind of changes seen in an embryo as it develops. Traditional German biology applies the same term to both forms of development: *Entwicklungsgeschichte,* which literally means "development history." Indeed, Unger envisaged that one species might change into another by a kind of "metamorphosis," comparable to the metamorphosis of a caterpillar into a butterfly. The notion that living creatures unfold the way they do because of their innate properties—like a bud unfolding to become a flower—is also reminiscent of Hegel's view of human history. But Darwin was keen to emphasize that his own vision of "descent with modification" was quite distinct from the processes of embryonic development. An embryo is "programmed" (as we would say today) to develop along a specific path, while according to Darwin's idea, natural selection could prompt each and every lineage to change in any number of ways. A bear, said Darwin, might become a whale. Because he wanted to avoid confusion, Darwin more or less avoided using the word *evolution* at all, let alone *Entwicklungsgeschichte.* So Unger's pre-Darwinian view of evolution was quite different from Darwin's. Nevertheless, Unger maintained that creatures do indeed change through time. Evolution *in general* was in the air—at least in some places—before Darwin wrote *Origin;* although only Darwin, and a few other visionaries, hit on a plausible mechanism: natural selection.

Finally, Unger was a populist, another widespread tradition of mid-nineteenth-century scientists, manifest in Britain in Michael Faraday, John Tyndall, and Thomas Henry Huxley. Unger wrote weekly Botanical Letters in the local newspaper, *Wiener Zeitung* ("Vienna Times"), and in 1852 he wrote that "botany . . . will come to be considered the physics of the plant organism," while the plant itself was a "chemical laboratory, the most ingenious arrangement for the play of physical forces." Unger knew too, in his bones, that there is indeed order in inheritance, and that one day that order would be illuminated by applying the methodical thought processes of the physicist.

But as always, anyone who thought clearly and radically in the Austro-Hungarian Empire was liable to be slapped down. The editor of the *Wiener Kirchenzeitung* ("Vienna Church Times") proclaimed that Unger's evolutionary views were "scandalous." The University of Vienna was Catholic, and moves were made to dismiss Unger. He was saved by the minister of education, Graf Leo Thun.

Still, Mendel's passion at Vienna was physics. He wanted to show how the workings of the Universe can all be traced to "a small number of laws," laws that should, he felt, be expressed in the language of math. Of course in his later studies of heredity, though not of the Universe at large, he did just this. He did indeed reveal "a small number of laws" (although these were formally stated not by Mendel but by later writers, based on his results), and his approach was unremittingly mathematical. Many biologists today believe that biology can make robust progress only by applying math, though the naturalists remain as essential as they always have been because only they can point out the problems in the first place. Mathematics plays a crucial part in modern evolutionary theory, not least in the guise of game theory, and the modern methods of classification, based on the techniques of cladistics, cannot be carried out on a serious level without the statistical analyses of computers. Mendel might be acknowledged not simply as the father of genetics, but as one of the first biologists to show how math can be used to solve the problems of life, just as it does the problems of the physical Universe at large.

MONK, TEACHER, SCIENTIST

Mendel finished at Vienna and returned to the monastery at Brno in 1853. He might have repeated the teacher's exams at that time, but he did not do so for another two years. No one knows why. Neither did he take a doctorate.

He did, however, probably the following year, begin his experiments with peas. There is no direct reference to this—there is a gap in the records—but there is a reference to work in progress in 1854. Eighteen fifty-four, then, was another key year in Mendel's life, and in the history of science and of humankind.

How he found time for his experiments, however, is another mystery. In 1853 he also went to work at the *Realschule* that had opened just a couple of years earlier, where he taught 18 to 27 lessons of physics and natural science each week to classes of between 62 and 109 pupils. He was (so his pupils recalled when biographers inquired of them after 1900) a fine teacher, with "blue eyes twinkling in the friendliest fashion through his gold-rimmed glasses." He used to invite his pupils into the monastery garden; one visitor from 1856 remembered that Mendel showed him "the plants with which he was doing crossing experiments, and his beehives," although, to be sure, it is not clear that there were any beehives at that point. Mendel did not acquire any until 1870. Mendel is also said to have explained artificial fertilization to his classes, and he led botanical excursions into the countryside around Brno and dug up hawkweeds (the *Hieracium,* which were to plague him) to bring back to the garden.

The *Realschule* had a distinguished teaching staff, which formed the core of the Natural Science Section that was established in 1849 within the Agricultural Society. Among them was the outstanding physicist A. Zawadski (1798–1868), who also stressed the need for organized research. Zawadski had a particular interest in botany, zoology, paleontology, and meteorology and wrote, in 1854, that we must ". . . learn to see sagaciously, think correctly, and feel warmly and deeply." Again, this deliberate engagement of thinking and feeling was very much a part of the tradition of *Naturphilosophie.* In that same year Zawadski was elected secretary to the committee of the Natural Science Section. That year, too, he gave a lecture entitled "The Developmental History of Lower Animals, namely on the copulation or blending of two animals to a single one for the purposes of fertilisation," which was a crucial theme for Mendel. At that 1854 meeting Mendel was accepted as a full member of the section, on Zawadski's recommendation. For Mendel, Zawadski was yet another excellent contact.

Meanwhile, however, the political life of the monastery, and indeed that of Europe as a whole, remained precarious. After 1848, the forces of conservatism again raised their heads, including the more obdurate wing of the Catholic Church, which, in 1855, reached a concordat with the state. The ecclesiastical authorities began a campaign to restore the spiritual life of the monasteries. Orders came from the Vatican via the archbishop of Prague to the bishop of Brno, A. E. Schaffgotsche, to visit the Augustinian monasteries of Bohemia and Moravia. He visited Brno in June 1854 and interviewed all the members of the community. In September he reported to the archbishop that Abbot Napp had committed many public offenses and was unable to devote sufficient attention to his main responsibility, which was, of course, to run the monastery. In fact, said Schaffgotsche, "the last ray of spiritual life" had faded away.

The monastery suited Napp and Mendel as it was, of course, but the bishop saw their interest in science and teaching as a contradiction to their spiritual calling. The list of enormities from this turbulent monastery was impressive. Nedele had been prevented from teaching because of his "rationalist" approach to the Bible. Klacel had been dismissed as a teacher for his "pantheist fantasies." One monk named Rambousek had been seen bathing in a public place "almost naked." Mendel had studied science in a secular institution so as to teach it at a *Realschule*. Yet, said the outraged Schaffgotsche, none of the monks would admit the error of their ways. Instead, they stood behind their abbot, and even had the gall to ask for a change of rules so that they could spend more time in teaching and science! The pope, said the bishop, should dissolve the monastery. Napp should be pensioned off and the rest of the monks dealt with individually.

Napp, however, was not so easily disposed of. He sent a memorandum to the archbishop proposing that the monastery be linked to the *Chorherren* Augustinians, which would allow the monks to continue studying science and teaching it. Like any good lawyer, he appealed to precedent: Pope Sixtus IV had granted comparable exceptions to the Augustinians in 1484. The archbishop acted reasonably swiftly. In 1855 he wrote to all the monasteries, listing their shortcomings. But he said nothing about dissolving Brno. Napp, for his part, agreed to repair the faults, and he and the monastery survived.

In 1855, Mendel went back to Vienna to take the teacher's exams that he could presumably have taken in 1853. Evidently he was nervous, and perhaps ill. In any case, he fell foul of another examiner, professor E. Fenzl, who, incidentally, was the grandfather of E. von Tschermak, one of the biologists who rediscovered Mendel's genetics research.

To understand Fenzl's objections we must anticipate a thought from the next chapter.

In the mid–nineteenth century it was not entirely clear that both parents in a sexual union contributed to the offspring, and it was still not accepted universally that plants were sexual beings at all. Some—known as spermatists—maintained that only the male contributed hereditary material to the offspring. Fenzl was, in effect, a spermatist. At least, he maintained that the plant embryo arose from the pollen tube and that the female cells merely provided nourishment. Mendel, of course, maintained that both parents contributed equally, and he made much of this in his published paper of 1866, which contains a definite note of triumph. For the moment, however, Mendel failed his teacher's exams yet again, apparently because of this very point. So he experienced yet another of his mental and physical crises. He returned to the *Realschule* as well respected as ever, but without the teacher's certificate he was ranked as a substitute teacher, and for the rest of his career he had to make do on half pay. He also returned to his precious rows of peas as soon as he got back from Vienna, and, as Orel says, "he was helping to unravel the very problem which may have led to his examination failure."

Mendel was active in Brno's Natural Science Society, which achieved independence from the Agricultural Society in 1860, albeit in the face of considerable protest from the latter. He was present at the meeting in December 1859 when Zawadski proposed the breakaway. Napp welcomed the new society, but said he hoped they would continue work on agriculture. Count von Mittrowsky was elected chairman, Zawadski became deputy chairman, while Dr. Schwippel, who taught natural history at the Brno *Gymnasium,* became secretary. Johann Nave was treasurer: he was a pupil of Unger's who had demonstrated the process of fertilization in algae, a key discovery. There were 142 founding members, including Mendel, and the number rose to 171 in the following year. The 24 honorary members included Bunsen, Unger, Virchow,

Wohler, and Purkinje, a distinguished lineup indeed. Mendel, in short, was very well connected. The new Natural Science Society finally achieved autonomy at the end of 1861 and held its first working meeting in January 1862. Plant hybridization was on the agenda. In February 1865 and again in March, Mendel delivered two lectures on plant hybridization to the monthly meeting. He published his one significant paper, based on those lectures, in 1866.

One last, intriguing detail from this middle and main stage of his life. On top of all the teaching, and the work on peas at the monastery, Mendel looked after the natural science collection at the *Realschule.* So it was that he helped to prepare wall displays on crystallography that were sent to the Great Exhibition in London of 1851. Mendel himself visited London a year later on a whirlwind tour in which, between July 24 and mid-August, he and his companions also took in Vienna, Salzburg, Munich, Stuttgart, Karlsruhe, Strasbourg, and Paris. Many have wondered whether Mendel visited Darwin during his London visit, but there is no evidence that he did; and since Mendel did not speak English, nor Darwin German, it is not clear what such a meeting would have produced. Surely, too, the modest Mendel would have been far too diffident to visit the great man (even if he had had time for the journey out to Kent).

Overall, however, it is evident that Mendel was extremely well connected. He delivered lectures on his work to what was surely one of the most distinguished societies of scientists in Europe. He also published. Yet no one seems to have registered what he was up to. Abbot Napp was his great champion and patron—he had a greenhouse built especially for Mendel—but even he does not seem to have appreciated his own protégé. Thus he commented in 1862 that hybridization was still "a problem for science," yet by this time Mendel's work was well in train. Why did he apparently not realize that Mendel had already solved most of the problems, at least as they appeared at the time?

THE LAST YEARS AND AFTER

We will leave his biography here. The background science—the problems Mendel set out to solve in his experiments with peas, and the conceptual difficulties that lay in the way—is discussed in the next chapter. The experiments themselves are discussed in chapter 4. After 1866, however, he did nothing of lasting scientific significance. Abbot Napp died in 1867, and Mendel took over as head of the monastery. He was not the first choice, but the man who was, the brilliant Tomas Bratranek (1815–84), had gone to be a professor of literature at the University of Kraków. As abbot, Mendel spent much of his time locked in dispute with the Catholic bureaucracy.

He retained his interest in science, however, and he attempted to take his work on peas to the next logical stage. Mendel's method, and his strength, was to identify problems that he knew were probably soluble and then move step by step into more difficult areas. He selected peas for study because he knew that of all the plants he might have looked at, peas were likely to give the clearest and most informative results. But he was also aware of the weakness of this approach: that nature cannot be second-guessed, and that it cannot simply be assumed that rules discovered in very simple cases can necessarily be applied to more complex ones. He sought to pin down the universal laws of heredity, if such existed; and he knew that to do this, at some stage he had to look at a wide variety of creatures. So after the work on peas, he looked at patterns of heredity in at least 17 other species. But apart from what he called some minor experiments with *Phaseolus* beans, all his later research remained unfinished. The climax was his superb research into bees—technically and conceptually quite brilliant—which took him up to the time of his death.

Mendel did, however, spend an inordinate amount of time—several years—trying to make sense of the hereditary patterns in the hawkweed *Hieracium*. He had no success. Now the reason is known. *Hieracium,* like some other members of the daisy family Compositae (including the dandelion), is parthenogenic. That is, the next generation arises from an *un*fertilized egg, although to add further confusion, in *Hieracium,* contact with pollen is required to trigger development. In the end, though, only the female genes are passed on to the offspring. In peas, like most other plants and animals, both parents contribute to the offspring. Thus the patterns of inheritance Mendel saw in peas simply do not occur in hawkweeds.

In his paper of 1866, Mendel shows that he had a knack for identifying such intractable problems and steering around them. His instinct surely told him that there was something peculiar and for the time being incomprehensible about *Hieracium,* and he should surely have realized that persistence was a waste of time. Good scientists know when to retreat and fight another day. So why did he waste so much effort on this obvious anomaly?

The answer is that he was encouraged to do so by the Swiss botanist Carl Naegeli, with whom he corresponded for several years after 1867. This correspondence was made known to the world in 1905 by Carl Correns, one of the scientists who unearthed Mendel's 1866 paper. It reveals Mendel's rising frustration, and tells us a great deal more about his ideas. It is abundantly obvious, however, that Naegeli missed the point. He simply was not paying attention.

Mendel was not a fool: far from it. Neither was he a wimp, as he showed in his many political forays. So why did he continue this correspondence with the astonishingly perverse Naegeli? Many have suggested that Mendel simply did not know anyone else of comparable stature, but this explanation will just not do. Napp was an outstanding figure and virtually the patron of Mendel's work. He lived until the pea research was finished, although he did not seem to have realized its full significance. Through Mendel's membership in the Brno Natural Science Society, he knew just about everybody, including the great Rudolf

Virchow and the incomparable Johannes Purkinje, both honorary members at Brno. Through Unger, he was also connected to Schleiden, who lived until 1881 and so was around for almost all of Mendel's life. Unger himself seems to have been a much more adventurous thinker than Naegeli, so why not talk to him? Johann Nave was a friend of Mendel's. Another was A. Makowsky, who taught at the *Realschule* from 1858 to 1859, published on zoology and botany, and visited Mendel in 1865 to discuss ideas, when the work on peas was already more or less complete. Then there was J. Kalmus, a student of Purkinje's, and C. Theimer, who studied hybrids and presided at meetings of the Natural Science Society in 1865 and 1868, when Mendel talked about crossing peas and hawkweeds. All these were obvious confidants, people to bounce ideas off, just as Darwin did with Joseph Hooker and Thomas Henry Huxley. In short, Mendel knew many excellent scientists and was only a step away from the greatest. Why, apparently, did none of them see the significance of his work? Why did he waste so much time on Naegeli?

Well, in his time Naegeli was highly regarded, and among his most fervent admirers was Unger himself. Naegeli had also studied with Schleiden, in 1842 (at Jena). He had done excellent work in plant physiology. He had shown how plants grow: how the cells of a plant divide at the apex of the shoot and roots. This might not seem relevant to studies of heredity, but plant science was much more all-of-a-piece in the mid–nineteenth century than it is now. Naegeli's was exciting work, showing how the common, everyday phenomena of plant growth are underpinned by the maneuverings of cells. Unger predicted that Naegeli would penetrate "deep into that labyrinth which no eye has yet seen" (although, as Orel comments, it was Mendel who truly did this). To Unger, indeed, the future seemed to belong to Naegeli. Perhaps Unger directed Mendel toward Naegeli. Probably Mendel felt that he could not abandon Naegeli without offending Unger.

So perhaps, in the end, it was Mendel's natural politeness, his peasant diffidence, and the fiercely hierarchical structure of German science that foiled him. Some men were perceived to be great—Virchow, Schleiden, even Naegeli—and some were not. Mendel, for whatever reason, was not. He hadn't got a doctorate; he had failed his teacher's exams. Nowadays, at least in Britain and America, young scientists are expected to be cheeky. They have to fight their corner. The young James Watson, who together with Francis Crick worked out the three-dimensional structure of DNA, is the modern archetype. But in nineteenth-century Germany, young scientists without formal professional status were expected to defer, and that, apparently, is what Mendel did. Perhaps in the end nobody took any notice simply because, nice fellow that he was, he had no professional or social standing. He in turn seems to have sacrificed his career for a social nicety. Stranger things have happened.

But although Mendel's extraordinary work was forgotten for a time, he personally was not. As abbot, he was a member of the Moravian Diet and attended many a meeting on local social matters. He was a popular teacher who stayed in touch with his pupils. So those who tried to reconstruct his life and work—in the early days including Correns (1900), Tschermak (1901), and Bateson (1902)—did find people who remembered him. But they remembered, mainly, his institutional work and personal character. Those who helped with the recovery process included Mendel's nephews, Alois and Ferdinand Schindler, whom Mendel had assisted through their education. Alois Schindler gave a lecture on Mendel in 1902, when the first memorial plaque was unveiled, at the fire station at Hyncice. It mainly commemorates Mendel's initiative in raising money for the station's equipment. Nonetheless—there was so little to go on—Bateson drew on this speech when writing a brief life of Mendel in a paper in 1902. Bateson also visited Brno in 1904 to find out more, but he could find no record of Mendel's experiments. His visit roused the procurator of the Augustinian monastery, A. Matousek, to collect such documents as referred to Mendel and to build a small Mendel museum.

The man who did most to restore the world's knowledge of Mendel was his biographer Iltis (1882–1952), who, among other things, found documents relating to his time at Vienna University. Yet in 1943 another Brno scientist, O. Richter, criticized Iltis for presenting Mendel as a free-thinking scientist and an admirer of Darwin. Richter, in contrast, underlined Mendel's qualities as a cleric. Richter's paper was published during the Nazi occupation of what was then called Czechoslovakia.

By 1948, however, Stalin's Communists were firmly installed there, and Stalinist biology was dominated by Trofim Lysenko, who rejected Mendel's ideas on political grounds. Mendel had shown, after all, that the way an organism turns out is very strongly influenced by its parentage: the begotten resemble their begetters. But the Communist state, with its simplified and coarsened version of Marx, needed to argue that people can be shaped by their environment, which to the Communists meant by the will of the government. The Mendelian notion that organisms—including people—were restrained by their biological inheritance was an anathema. In a series of crude experiments, Lysenko contrived to show that wheat is just as amenable to environmental manipulation as people are. Effectively, he contrived to show that wheat could survive Russian winters if given a sharp talking to. It couldn't, of course, and under his guidance the harvests failed.

The "conflict" between the long-dead Mendel and the all-too-extant Lysenko was a grotesque example of the "nature versus nurture" debate that has rumbled through biology and philosophy for centuries. Lysenko and Stalin between them demonstrate that "environmental determinism" can be at least as horrible as any kind of "biological determinism." In truth, as Mendel well knew, organisms are a dialogue between their "nature" (underpinned by their genes) and their "nurture" (their environment). Lysenko's insistence that environment is all cost the lives of many thousands of Russians, as their crops perished in the snow. It also showed how badly science and politics mix, and that, in the end, politicians are obliged to work within the confines of the real Universe; they cannot simply decree nature to be different, in the way that King Canute once tried to rule the tides (although Canute, unlike Lysenko, was expressly trying to show that nature could not be overruled). Lysenko's nonsense also encouraged Stalin to ban the otherwise burgeoning science of genetics. As a consequence, the Mendel museum was closed in 1950, and most of the Czech monasteries were closed at the same time, so the museum was caught in two ways. Fortunately, most of the documents and instruments from Mendel's time were rescued.

Iltis himself went to the United States before World War II, taking copies of documents relating to Mendel that are now housed in the Museum of Natural History at the University of Illinois at Urbana-Champaign, which may not be particularly well known in the world at large but has produced some fine biology. In 1965 the Mendelianum in the Moravian Museum in Brno published 120 documents and photographs relating to Mendel—100 years after he published his paper.

So that, in outline, is Mendel's biography: a bizarre tale that somehow manages to encapsulate much of Europe's history: the end of feudalism, the rise of modern agriculture and industry, science as philosophy, science as the generator and handmaiden of technologies, science versus religion, science versus state, ideology versus ideology, the rise and fall of dictatorships. Mendel came of "peasant stock," but he was not a country bumpkin, or any kind of primitive. His family was supportive; he was highly educated and moved among the greatest scientists of Europe. Science in general was developing apace, on deep (if sometimes confusing, to modern eyes) philosophical foundations. The philosophy of science included the powerful notion that the universe as a whole is guided by laws, and that these laws can be expressed mathematically; and at least some biologists were beginning to realize that the rigor of physics and math should be applied to the problems of living creatures, as well as to the physical world. Many scientists, too, including Mendel's own teachers—Doppler, Ettinghausen, Unger—were stressing the supreme importance of elegant experiment. Mendel took all this in. At the same time, he was steeped in agriculture: through his own family, through society at large, and through the particular inclinations of his monastery at Brno.

In short, he seems to have been made for the task. How that task appeared in the mid–nineteenth century, we will explore in the next chapter.

THREE

Breeders, Scientists, and Philosophers

Isaac Newton said in the seventeenth century that if he had seen further into the laws of nature than others, it was because he had "stood on the shoulders of giants." Mendel could have said the same when he began his work with peas in the mid-1850s. There had been key advances in the philosophy of science, with increasing stress on the need to quantify and to carry out experiments with a chance of producing a solution. This chapter looks more closely at the new ideas in science itself and—equally important—in the craft of breeding. But in the mid–nineteenth century the ideas that are now known to be accurate were mixed in more or less equal measure with notions that were simply wrong, or hopelessly confused, and yet were argued with equal vigor. These confusions should also be examined, to show the emerging truths in proper perspective.

But what was the problem? At first sight heredity seems perfectly straightforward. Like begets like. Cats have kittens, women have babies, oak trees spatter the woods with acorns that become, miraculously, a new generation of oak trees. It is, though, as Niels Bohr said about particle physics: if you think it's easy, you haven't understood the problem. To begin with, the whole business of heredity has several threads to it. How, first of all, does any creature actually produce offspring at all? What are the mechanics of reproduction? This in turn has at least two discrete components: how parents make an embryo, and how the embryo grows and develops to become the finished creature. Then there is the more abstract matter: How do parents ensure that the offspring does indeed resemble them? What is the nature of the "information" that passes from generation to generation?

Now that we understand each of these threads—up to a point—we can tease them apart. They are studied in separate university departments. How eggs and sperm are made, and come together, and how the resulting embryo is succored: these are the substance of Reproductive Physiology. How the embryo grows and develops: that is Embryology, or Developmental Biology. How hereditary information is passed from generation to generation: that is Genetics. But before the details were understood, these three areas of study were commonly conflated. This was far from stupid. Nature *might,* for example, make copies of creatures by taking a fragment from each part of the parents' bodies and cobbling them all together to form a new individual. In such a case, the business of reproduction and the business of heredity would effectively be the same. There would be no genes providing instructions on how to make flesh: the flesh itself might be passed on. That is not how things happen, as it turns out; at least not in sexual reproduction, as practiced by human beings (although it is similar in some ways to the asexual reproductive strategy of corals and strawberries). But that *might* have been the mechanism. In the absence of better information, it seemed perfectly reasonable to study reproductive physiology, embryology, and heredity all together. Such conflation, however, did not make life any less confusing.

Let us ask how "like begets like," as the problem still appeared in the mid–nineteenth century, and thus as Mendel confronted it. Not until we perceive the problems can we appreciate the answer and be dazzled, not disappointed, by its simplicity.

WHO CONTRIBUTES WHAT?

Consider first of all what we now take for granted. All creatures are divided into two sexes: male and female. Both contribute to the offspring, and indeed, barring a few details that need not delay us for the moment, they contribute equally. Typically, the female provides a (relatively) big, yolky egg, and the male provides a spermatozoon, or some variation thereof: always tiny, usually motile, and packed with hereditary information. The two "gametes" (female egg and male sperm) combine to form a one-celled embryo (technically known as a zygote), which divides and divides again in careful choreography to form an embryo, which grows and unfolds to become a facsimile, or near facsimile, of its parents, combining features of both. In these enlightened days of sex education, every 7-year-old could tell you all this. Plants do the same kinds of thing in principle, although in seed plants the male gamete is cocooned within the protective pollen.

But—if we didn't have the benefit of 400 years of science—how much of this *should* be taken for granted? Almost none. At least, we could be sure that most of the creatures we are most familiar with, including human beings, are divided into male and female, and that both are indeed required for reproduction; and (in the creatures we are most familiar with) that the males must impregnate the females with semen, which is conspicuous enough. But then the difficulties begin.

Is it really obvious, for example, that both parents contribute to the offspring? Even if they do, *what* do they contribute? Some scientists, in not-so-ancient times, proposed that the sperm alone provided the hereditary information for the next generation and that the egg—with its yolk—merely provided nourishment; a hamper for life's journey. Others averred with equal vehemence that the egg provided all that was really necessary and that the sperm merely triggered development. The former were "spermists," the latter "ovists." Of course, before microscopes were developed in the seventeenth century, nobody had seen a sperm—they had seen the seminal fluid, but not the all-important whirling cells within—and neither had they seen a mammalian egg; so before the microscope, this debate was conducted in more abstract terms.

Surely the microscope would resolve this issue? In seventeenth-century Holland, Anton van Leeuwenhoek (1632–1723), who made hundreds of microscopes in his long life and is generally acknowledged to be the first great microscopist, saw sperm in his own semen and duly became a spermist. Meanwhile, in Italy, Marcello Malpighi (1628–94) observed an entity in an unfertilized bird's egg that he took to be the germ of the next generation, and became an ovist. The spermists and ovists may sound, from this twenty-first-century vantage point, like Jonathan Swift's Little Enders and Big Enders, who attacked their boiled eggs from different ends and sought to resolve the matter by going to war. But Leeuwenhoek and Malpighi were great biologists. The question they were asking—who donates what?—just wasn't easy. Anyway, even if both parents do contribute hereditary material to the offspring, should we assume that they contribute equally? If so, why do children sometimes resemble one parent more than the other? Why is a son, sometimes, the "spitting image" of his mother, or the daughter of her father? Might it be that on different occasions, each parent might contribute a different amount?

In the end—as is so often the case—the reality has proved more complex than anyone might have supposed. The sperm does indeed provide hereditary material (what we now call genes), but it *also* triggers development, in the way the ovists suggested, in a process known as activation. In some plants, the pollen may activate the egg *without* contributing genes. This is true of dandelions and hawkweeds, which is why Mendel could make no sense of hawkweeds when Naegeli persuaded him to study them. In all known cases, the egg does of course provide hereditary information in the form of genes, but it also provides nourishment in the form of yolk. This is true even in mammals, before the embryo implants in the wall of the uterus.

Whether we are spermists or ovists or concede that both gametes contribute, we must ask: What form does the hereditary material take? Here again there were two clear schools. The "preformationists" believed that the sperm (or the egg) contained a complete little replica of the offspring. There are seventeenth-century pictures of spermatozoa with little fetuses coiled up in the head. Laurence Sterne lampoons this notion in *Tristram Shandy* as he muses in his freewheeling way on whether onanists commit mass murder as they scatter their seed upon the ground, each one a primordial human being. Perhaps some microscopist saw such a creature: the power of wishful thinking is wonderful. Leeuwenhoek described spermatozoa as *Samentierchen:* "little animals," or "animalcules," as the expression is often quaintly rendered. Others maintained that the egg, or the sperm, or both, merely contributed something in the nature of modeling clay, to be fashioned into an embryo. This was the notion of "epigenesis." Aristotle, writing in the fifth and fourth centuries B.C., was the first known author to suggest that both parents contribute to the hereditary makeup of the offspring, and he perforce was an epigenicist. He envisaged that the male contribution of "modeling clay" took the form of semen, while the female's contribution was contained in the menstrual blood. Semen was also derived from blood, he thought, from "sanguineous nutriment." This is a quaint idea, yet in the absence of better information, it is perfectly reasonable.

In the early nineteenth century new ideas arose that truly began to make it possible to think in modern terms. In particular, the further development of the microscope gave rise to "cell theory": the idea that big creatures like animals and plants are compounded from many tiny cells. The rise of cell theory is an example of what, in the 1960s, the American philosopher of science Thomas Kuhn called a paradigm shift: a sea change in the way scientists look at the world. In reality, said Kuhn, *all* significant progress in science is made in jerks, in such "paradigm shifts."

Malpighi set biology on the road to cell theory. At least, he showed the existence of the capillaries, the fine blood vessels in the body tissues that run between the arteries and the veins, the discovery that rounded off William Harvey's vision of blood circulation. But the final insight had to wait until the 1830s. In 1837 the great Czech biologist Johannes Evangelista Purkinje (1787–1869) observed a general parallel between the structure of animal and plant tissue, both of which seemed compartmentalized. Then, in 1838, the German botanist Matthias Schleiden (1804–81) proposed that all plant tissue is in fact composed of nucleated cells; and in the following year his compatriot Theodor Schwann (1810–82) suggested that plant *and* animal tissue was compounded of cells. Schwann proposed in *Mikroskopische Untersuchungen* ("Microscopical Researches") that "cellular formation" might be "a universal principle for the formation of organic substances." You may recall that Purkinje was an honorary member of the Natural Science Society at Brno, of which Mendel was a founding member. He visited the monastery at Brno in 1850, and Mendel's close friend, Klacel, took him on a conducted tour. Schleiden, you will remember, taught Unger, who was Mendel's teacher in Vienna. Mendel indeed was well connected.

Schwann suggested that the individual cells simply condensed out of "nutrient liquid," like crystals of salt precipitating out of a rock pool, but this misconception was soon put to rights. In 1841 the Polish-German anatomist Robert Remak (1815–65) described how cells divide, and in 1855 the German pathologist Rudolf Carl Virchow (1821–1902) coined the dogma *omnis cellula e cellula*—"all cells come from cells." In other words, each of the vast battalions of cells of which a mouse or a sheep or a human being is composed (or an oak tree or a moss or a mushroom) derives from a single, initial cell, dividing and redividing. The German role in this thread of life science is obvious: Schleiden,

Schwann, Remak, Virchow—and German-speaking Mendel was close to the action. Virchow, too, was an honorary member of the Natural Science Society of Brno.

How, though, does the single initial cell—the fertilized egg—multiply to form an entire multicellular organism, like a sheep or a human being? What happens? What are the underlying mechanisms? Alongside German-based cytology grew German-based embryology. At first this new discipline was perceived as a branch of anatomy, but it was established as a specialty in its own right by the German-Estonian Karl Ernst von Baer (1792–1876). In 1827, in *De ovi mammalium et hominis genesi* ("On the Origin of the Mammalian and Human Ovum"), von Baer described how all mammals, including human beings, develop from eggs. He showed, too, how the different organs form as the embryo develops: in which order they arise, and from which tissues. He also established comparative embryology, which reveals the similarities and differences in the development of different creatures.

How, though, is the zygote formed in the first place? This is the key issue. This is the point at which—presumably—the parents pass on their inheritance to the offspring. The answer came straight out of cell theory, and again was revealed in the 1830s. In 1839 one M. Barry described how he had observed a spermatozoan penetrating a mammalian egg. This was a truly remarkable finding; although perhaps it was too remarkable, for many people disputed it. Purkinje accepted the observation, however, and drew it to the attention of his friend the physiologist R. Wagner.

Then, in 1855, the German botanist Nathanael Pringsheim (1823–94) showed the same phenomenon in *Oedogonium,* a freshwater alga. Algae are primitive plants that produce motile gametes—sperm—rather than pollen, as in flowering plants. Mendel had begun his own experiments at this point, and in 1858 his friend Johannes Nave lectured on Pringsheim's work at the Brno Natural Science Society. Again, Mendel was in touch with the developments.

So the most important elements of cell theory were established before Mendel began his great work on breeding in the 1850s, or soon after. Surely, then, most of the earlier confusions must already have been cleared up? Sperm and egg were known to be cells. They had been seen to fuse, both in animals and plants. The one-celled embryos to which they gave rise had been shown to divide, and divide again, to form the finished multicelled creature. There could be no more confusion.

Could there? Nowadays we've all seen eggs and sperm combining on many a TV documentary. Students observe this under the microscope. In vitro fertilization (IVF) is almost commonplace. But was Barry's report of fertilization in mammals—the observation of one biologist, with a shiny nineteenth-century brass microscope—really enough to carry the day? Closed-mindedness is the bane of science, as of every other human pursuit; but it is certainly proper to be skeptical. Besides, even if a new life did begin with the fusion of egg and sperm, what would that really demonstrate? Eggs are much bigger than spermatozoa, after all, so they might in reality contribute more to the offspring. On the other hand, they might simply provide nourishment, just as the spermists maintained. Or the sperm might simply be triggering development, as the ovists thought. The general lesson is that in science, one observation does not shift a paradigm all by itself—and indeed it should not.

Mendel failed his final attempt to qualify as a teacher in 1855, when he came up against an essentially spermist examiner, E. Fenzl. Mendel's dissertation included a passage on the fertilization of flowers. By then the cellular process was known: after the pollen lands on the stigma, it sends out a "pollen tube" that runs down the style into the ovary and fuses with the ovum. But Fenzl maintained that the new embryo grew entirely from the pollen tube, that the female cells in the ovary merely provided succor. Fenzl was clearly a diehard—his obduracy shows nonetheless that the ancient arguments were by no means resolved.

These, then, are some of the uncertainties that surrounded even the most basic mechanism: the fusion of a male sperm with a female egg to create a zygote that contains elements of both. If the mechanics were different, then the rules and the patterns of heredity would also be different. Presumably the patterns of inheritance would be less consistent than they are if the two parents contributed differing amounts to the offspring on different occasions. But then, as all breeders of crops and livestock knew full well—and indeed, as all human beings know who have ever looked at their fellows and at other living creatures—the patterns of inheritance seem extremely inconsistent even as things are.

LIKE BEGETS LIKE—BUT ONLY UP TO A POINT

To be sure, women produce baby humans, and cats produce baby cats. But although the offspring do resemble their parents and each other, the resemblance is never exact, although it may be almost exact in the case of identical twins. Variation is an obvious fact, and one that requires explanation.

Of course, when two animals or plants that are slightly different are crossbred—or "crossed," or "hybridized"—we expect the resulting "hybrids" to vary, both from each other and from the parents. But sometimes two parents that look the same—at least with respect to particular characters—also produce offspring that differ from themselves and each other. For example, a man and woman who both have brown eyes might produce a child with blue eyes. Worse: the quality of blue-eyedness might disappear from a family of brown-eyed people for generations at a stretch and then suddenly pop up. The overall distribution of blue-eyedness is, as biologists tend to say, "spotty." Your grandparents might all have brown eyes, but your great-grandmother, your Aunt Gertie, and your cousin George have blue eyes. We need not assume double-dealing, or evict the blue-eyed lodger. That's just the way heredity works.

Similarly, two blue-flowered garden lupins when crossed are liable to produce offspring with flowers of many different colors. Unexpected variation in color is one of the attractions of traditional lupins, resplendent in their cottage gardens. On the other hand, the offspring of two blue-eyed human beings will *always* have blue eyes, and the offspring of two white-flowered garden peas will *always* have white flowers. Yellow-flowered garden peas, however, may produce at least some white-flowered offspring, just as brown-eyed humans may have some blue-eyed offspring. But it is possible at least in theory to identify some "strains" or "lineages" of yellow-flowered garden peas that produce *only* yellow-flowered offspring, and some brown-eyed people who have *only* brown-eyed children. The lineages that consistently produce offspring with a particular character are said to be "true breeding."

Such variation, however, is only half of the matter. For sometimes, as all breeders know (and any large family of human beings is unfortunately likely to know), offspring are produced who look nothing like their parents or their siblings. Animals are sometimes born with strange deformities, and then were traditionally known as "sports." Children may be similarly afflicted and are sometimes born with peculiar diseases. Again, such pathologies may come about in various ways, which are obvious when the rules are known. When such anomalies seem simply to drop from the sky, however, they muddy already difficult waters even further.

Then again, *some* anomalies that are present at birth are indeed caused by the hereditary factors that we now call genes, but some are not. Some may be caused by accidents in the womb, and some by pathogens that are passed from generation to generation ("vertical transmission"), such as congenital syphilis, which features powerfully in nineteenth-century literature. Not all such heritable pathogens are unremittingly evil, however. In the seventeenth century the Dutch developed a craze for tulips that at least matched England's South Sea Bubble, and the present zeal for dot-com companies. Striped tulips were especially highly favored. Stripiness is brought about not by particular genes, but by viruses, transmitted from bulb to bulb. It is a temporary feature: as the generations pass, the infection takes hold, the plant is weakened, and the breeder must begin again. The seventeenth-century breeders did not know about viruses, but they knew that stripes were a strange character that followed its own rules, and they produced new lines of striped tulips not by the normal sexual methods but by exposing bulbs to infected lines. To the observer, however, stripiness is merely another "inherited" character.

Then, as every breeder knows—and as all human tribes reflect in their folklore—some matings and marriages have a good outcome, while others are much less satisfactory. Again the underlying rules are simple; but again, before the rules were known, the picture was eminently confusing. Thus, in general, a fine, lusty male animal with a good, robust female seem likely to produce exemplary offspring. Often, indeed, this is the case. But if the lusty male is related to the robust female—if indeed they are brother and sister—then their offspring could be a disaster. This would be an example of "inbreeding," which leads to what modern geneticists and traditional breeders alike call inbreeding depression. Most human tribes have powerful taboos against the incestuous matings that lead to inbreeding. Oedipus tore his eyes out when he realized he had inadvertently bedded his own mother. Aristocratic families that practice inbreeding to avoid contamination with commoners' "blood"—or, more to the point, to keep wealth and power in the family—are commonly plagued with diseases of an inherited nature, such as the hemophilia and porphyria that ran among the many relatives of Queen Victoria. Many a family of inbred aristocrats has simply died out. Before modern breeding protocols were developed, inbreeding put an end to most attempts to breed animals in zoos. Breeders, in short, are in a constant state of tension. They want to cross the best with the best. But the best individuals in any one population are quite likely to be closely related to one another. If relatives are mated, this is liable to lead, sooner or later, to inbreeding depression: offspring with many kinds of enfeeblement.

So what's the answer? Obviously, to mate creatures that have desirable qualities but are not closely related. Any kind of mating between nonrelatives can be called outbreeding. If the two potential parents belong to distinct "varieties" or "breeds," such outbreeding is called hybridization. Sometimes the "hybrids" produced by such matings are extremely fine, being stronger and more intelligent than either parent. Then they are said to partake of "hybrid vigor."

All this has been well known to breeders of crops and livestock for hundreds of years, and doubtless for thousands. Hybrid vigor is widely recognized in human societies, too. In literature, the offspring of interracial marriages tend to have disparaging names, like "mulatto," "half-breed," "half-caste," and "Cape-colored," implying "neither one thing nor the other." On the other hand, such "mulattoes" have often been disturbingly athletic and quick-witted. Similarly, the stable boy who bears an intriguing resemblance to the lord of the manor and is significantly more alert than any of the legitimate heirs is a stock character in many a pulp novel.

But although some hybridizations work very well, some do not. Matings between greyhounds and other breeds of hound produce wonderfully vigorous offspring known as lurchers. Many a working dog—gundog, sheepdog, for example—was traditionally produced by judicious crossing of "purebred" or "pedigree" parents of different breeds. But some interbreed hybridizations produce unfortunate results. St. Bernards and Great Danes are a sad mismatch. Nowadays the reason is clear. St. Bernards are big because they produce a great deal of growth hormone, which makes them generally heavy and floppy. Great Danes are big simply because they have genes that produce long limb bones. St. Bernards are *adapted* to a high output of growth hormone and are not distressed by it. They merely become large, if a bit lethargic. But the hybrid offspring of St. Bernards and Great Danes produce excess growth hormone yet lack the means to cope with it, and their general physiology is awry. In short, *some* exercises in outbreeding work very well, and lead to hybrid vigor; but some have a poor outcome, leading to "outbreeding depression."

This brings us to an issue that taxed breeders and philosophers—and theologians—alike for many centuries, and still causes confusion among nonbiologists, while even biologists continue to argue: the difference between "varieties" (or "breeds," or "races") and "species."

Again, at first sight, the issue seems simple enough. Few of us have any problem with the idea that St. Bernards, cocker spaniels, and Chihuahuas are all variations on a theme of domestic dog, different breeds of the same species, known to scientists as *Canis familiaris,* while wolves, though related, are of a different species, *Canis lupus.* Similarly, though somewhat more complicatedly, we can accept that cabbage, kale, cauliflower, and brussels sprouts are all variations on a theme of the wild cabbage *Brassica oleracea* (although, of course, each of those main categories—cabbage, kale, cauliflower, and sprouts—is further subdivided into the various seedsman's "varieties"), while the turnip, though related, belongs to a different species, *Brassica rapa.* The principle, at least, is easy enough to grasp.

On what, though, is this principle based? In Mendel's day, species were still defined largely by what they looked like. As Mendel said in his paper of 1866, by ". . . the strictest definition . . . only those individuals belong to a species which under precisely the same circumstances display similar characteristics." But he must have known that definitions based on appearance do not get us very far. St. Bernards and spaniels, for example, though casually classed within the same species, look as different as, say, a moose and a roe deer, which are clearly placed in different species.

On the other hand, some dogs—such as Alsatians and huskies—look very much like wolves; at least, their resemblance to wolves is much more striking than, say, to bulldogs. But again, huskies and bulldogs (like spaniels and St. Bernards) are placed in the same species. Yet some wild creatures that are placed in different species are almost impossible to tell apart. Chiffchaffs and willow warblers seem almost the same, apart from their song; this example would have been well known to mid-nineteenth-century European naturalists. Many more such pairings have come to light in the twentieth century. North American ornithologists now recognize a whole series of different owls that are almost identical and were formerly placed within one species. The owls recognize each other by their calls. Professor Simon Bearder, of Brook University in Oxford, studies bush babies, or galagos, in West Africa and seems to find new species on every visit. Again, the different types look much the same but recognize each other by their calls, and DNA studies typically show huge differences between them. Owls and bush babies are both nocturnal: to them, appearance doesn't matter, but voice does. Why, then, do modern biologists insist that two creatures that look very different may be of the same species, while others that look almost identical are of different species?

The important point has to do with reproductive barriers. The standard twentieth-century notion, generally credited to the great German-American zoologist Ernst Mayr, is that two creatures will generally be ascribed to the same species if they can mate together sexually to produce "fully viable" offspring. They are placed in separate species if their offspring are not fully viable. "Fully viable" is a deliberately generalized expression. In effect, it refers to a creature that can compete in the wild world at least as well as either of its parents: it can find food just as efficiently, fight off enemies, attract mates, and produce healthy young. Arab horses and English Thoroughbreds, two breeds within the domestic species *Equus caballus,* produce excellent and much-prized offspring, both lithe and tough. But when *Equus caballus* stallions mate with the mares of *Equus africanus,* the ass, the result is a mule: a tough animal, but sexually sterile. A sexually sterile creature is clearly not fully viable. If the two extant species of camel are crossed—*Camelus ferus* (or *bactrianus*), the two-humped Bactrian camel, with *Camelus drome-darius,* the one-humped Arabian camel, or dromedary—the female offspring are fertile, but the male offspring are sterile. If camels are mated with horses (perish the thought), there are no offspring at all. The boundaries are too great.

Again, all this seems clear enough. Matings between different varieties (or breeds or races) of the same species—known as *intra*-specific matings—produce fully viable offspring. Matings between different species—*inter*specific matings—at best produce offspring that are not fully viable, and indeed may be sexually sterile, but generally produce no offspring at all.

Yet there are complications, many of which would have been well known to mid-nineteenth-century breeders and biologists. For one thing, intraspecific matings (between varieties of the same species) do not always produce fully viable offspring. We have seen one example: the offspring of St. Bernards and Great Danes are a sorry sight. Horticulturalists, however, would have been much more concerned with a phenomenon that occurs in many plants, including apples and plums. Such species go to great lengths to avoid matings between individuals that are too closely related to each other; in other words (to put the matter anthropo-morphically), they seek to avoid inbreeding. To this end, they have developed "specific mating barriers," which prevent mating between two individuals of the same variety, or closely related varieties. To produce a crop, apple growers have to plant individuals from different but compatible varieties side by side. So here we have mating barriers *within* species.

On the other hand, various pairs of discrete species will mate perfectly well together if given the chance. Domestic dogs and wild wolves will mate, given the opportunity, and the offspring are often formidable, and perfectly fertile. The wild cat of Europe,

Felis silvestris, is disappearing as a distinct species because of excessive interbreeding with feral domestic cats, *Felis catus.* In zoos, tigers and lions have often mated, and the offspring are fertile. In the wild, we often find "hybrid zones" at the boundaries where the ranges of different species overlap. Carrion crows in general live south of a line that passes through Edinburgh, while hooded crows live north of that line. Where the two ranges meet, in Edinburgh itself, we find interspecific hybrids. Similarly, in a line running north-south through central Europe, we find a hybrid zone between the fire-bellied toads that live to the west and the yellow-bellied toads that live to the east.

In short, the standard twentieth-century view of a species works well enough for most purposes—much better than any definition based on appearance—but there are still conditional clauses. In practice, wolves do not mate with domestic dogs if other wolves are available. Wolves *prefer* wolves. Similarly, tigers do not mate with lions in the wild. Their ranges *do* overlap in the wild, or at least they sometimes did until recent centuries, for although tigers do not live in Africa, there were plenty of lions in Asia and Europe well into historical times, and they still live wild in the Gir Forest of India. But in the wild, each sticks to its own kind. As the South African biologist Hugh Patterson put the matter: animals of the same species recognize each other's mating signals and certainly prefer them to the signals of other species.

Admittedly, hooded crows and carrion crows do interbreed in the wild. But the hybrid offspring have not spread into the home ranges of either parent species, suggesting that the hybrids do not compete well either with the purebred hooded crows to the north or with the purebred carrion crows to the south. So by the standards of crows, the hybrids are not fully viable. In human prehistory, it's clear that our Cro-Magnon ancestors lived side by side with the Neanderthals in Europe for at least 10,000 years. We need not doubt that they *could* have interbred, and possibly did so from time to time. Yet their fossil remains suggest that the two kinds of human mostly kept to themselves. *They* recognized mating barriers even if we cannot easily see what they were. So although modern humans and Neanderthals *could* have mated perfectly well together, in practice it seems they did not. Since in practice they did not, they should be treated as separate species. The same applies to the difference between, say, huskies, which are *Canis familiaris,* and wolves, which are *Canis lupus.*

The biological complications are not quite exhausted, however. Sometimes individuals that clearly belong to different species do mate together to produce offspring that are not only fully viable but are in fact a new species: able to mate perfectly well with each other, but not able to mate successfully with either parent. We will look at the underlying mechanism for this in the next chapter. This phenomenon is quite common in plants, though rare in animals. Mid-nineteenth-century breeders would not have known how it occurred, but they might well have known that the rutabaga, placed in the species *Brassica napus,* is probably a hybrid between the cabbage *Brassica oleracea* and the turnip *Brassica rapa.* Rutabagas seem to have appeared, apparently spontaneously, in seventeenth-century Bohemia.

All of this discussion brings us to an issue that was of key significance in the mid–nineteenth century, with implications that ran from the practical craft of breeding on the one hand, to the deepest reaches of theology on the other. Are the distinctions between species absolute, or is a species merely a variety writ large? Is it possible to turn one species into another? Can any one species divide, to produce more than one daughter species?

To understand the full significance of this, we must explore the history of Christian orthodoxy, and appreciate that Christian orthodoxy underpinned the general worldview of the mid–nineteenth century. This was true even though many intellectuals, at that time, had been turned away from religion by the rationalism of the Enlightenment. In the Middle Ages, first the Arab world and then the West rediscovered the art and philosophy of the ancient

Greeks. Through the offices of medieval scholars, notably but not exclusively Thomas Aquinas in the thirteenth century, the ideas of Plato and Aristotle were essentially fused with those of Christianity to produce a new, intricate, and satisfying theology, one that was to last for centuries.

Locked up in this theological synthesis was the idea of species. Plato argued that everything that exists on this Earth is but a model, a poor reflection, of some ideal that exists in Heaven. He seems indeed to have believed that, somewhere on high, the ideal table or chair—or dog or cat—has literal existence. Each ideal represents an idea of God. This became the Christian orthodoxy. The idea implied that species are not merely biological constructs. The notion of the species becomes, essentially, sacred. Each one represents an idea of God.

There were biologists before Darwin who entertained the idea that creatures had evolved through time: they had not simply been created in their present forms, as the story of Genesis suggests. Many, however, were able to reconcile the idea of evolution with the narrative of Genesis. They suggested that God, in the beginning, had created a series of primordial creatures that did indeed change through time: essentially, each original primordial form unfolded, generation by generation, as a rosebud unfolds into a flower. But the different lineages remained separate. The French pre-Darwinian evolutionist Jean-Baptiste Lamarck argued in this way. Lamarck did *not* envisage, however, that any one lineage of creatures could branch, such that any one primordial creature could give rise to several different lineages. That would have implied that at some stage God had changed His mind.

But Charles Darwin, in his seminal *On the Origin of Species by Means of Natural Selection* of 1859, does argue that species may divide to form several different species. In short, he does not envisage that many different, separate lines of creatures simply change through time. He explicitly proposes instead that all creatures now on Earth—and all that are known to have lived in the past—

arose from one single ancestor, so that we are all part of one great family tree. Thus he argues emphatically that species barriers are not inviolable. Note, too, that the title of Darwin's book—*Origin of Species*—was more provocative than it seems today. Ernst Mayr records in *Toward a New Philosophy of Biology* that many of Darwin's critics objected more strongly to this particular idea than to his broad proposal that creatures do indeed evolve. Evolution in general could be made acceptable, but the splitting of species, the fragmentation of God's original ideas, was widely perceived as blasphemy. In short, the distinction between "species" and "variety," or "breed," was not just a technicality in the mid–nineteenth century. The implications ran very deep indeed.

Despite all this deep thought, however, and alongside it, ran the simple mild confusions that we find in any age. Many still felt in a vague way that species do drift cryptically, one into another. Some claimed, for example, that many forms of wayside vetch were somewhat degenerate derivatives of peas and lentils.

These, then, are the broad problems of heredity as they still loomed in the mid–nineteenth century. But let us look more closely at the mechanisms of heredity. Whether we conclude that both parents pass on information to the offspring, or that only one does so, whether or not we concede that this information is packaged within the gametes, we still must ask, what *form* does the information take?

WHAT KIND OF INFORMATION PASSES FROM PARENTS TO OFFSPRING?

In 1837, Mendel's mentor at Brno, Abbot Napp, asked in the context of breeding sheep, "What is inherited, and how?" In particular, he and his contemporaries wanted to know: Is the hereditary information poured from the parent into the gametes like a fluid? Or do the instructions come in discrete units, like particles?

Essentially, the fluid-versus-particle debate on heredity reflects a dichotomy that has run through all philosophy and science from earliest times. Nowadays it is manifest as the contrast between what computer and hi-fi aficionados call analog and digital technologies: the first envisages things (substances, information) as continuous entities, like plasticine; the latter envisages the components of the universe as being compounded from tiny particles. Philosophers in the digital-particle school include the fifth-century B.C. Greek philosopher Democritus, who argued that the universe was built from a few kinds of particle; Francis Bacon, who in the early seventeenth century conceived that the complexities of visible phenomena reflect the orderly maneuverings of subvisible particles (an idea that seems to presage modern chaos theory); Descartes, who felt that creatures inherited particles of information from their parents (about which, of course, he was right); Leibniz, who envisaged a universe compounded of "monads," which, as the modern physicist Julian Barbour argues, are remarkably prescient of the units of modern particle physics; Newton, who argued that light is compounded of particles, which are now called photons; and John Dalton, who in 1804 first proposed the atomic theory that is the basis of modern chemistry. Nowadays we have modern particle physics and computer science, whose codes are digital. Schleiden and Schwann's notion that all bodies are compounded of small cells is essentially digital in nature.

Analog suggests not particles, but a continuum. Thus traditional clocks are analog technology, showing the time by moving their hands continuously. And while Newton argued that light is particulate, Christiaan Huygens showed that it must be a wave, and modern physicists acknowledge that both points of view are correct. A common analog view of inheritance is that instructions are passed from parent(s) to offspring by fluids. As we have seen, Aristotle envisaged a coalescence of seminal fluid and menstrual blood. Darwin of course did not follow Aristotle in this, but he did for a time entertain the idea (certainly when he wrote the first edition of *Origin*) that heredity might operate as if by the mixing of inks. An artist mixes blue with yellow to make green; so might male "ink" be mixed with female within their shared offspring. Many characters in the offspring do seem to be a straightforward averaging of those of the parents, though the mixing cannot be too thorough, or brown-eyed parents could not give rise to blue-eyed children. However, a polemical Scottish engineer named Fleeming Jenkin delivered the coup de grâce to this particular conceit of Darwin's. If heredity did operate simply by averaging, he said, then natural selection could not possibly work. Any new and advantageous character that happened to crop up would soon be diluted away as the lucky individual mated with others who were less well endowed. In this case, Mendel's digital view of heredity was correct, while Darwin's analog view was mistaken.

Whatever form the hereditary material took—particle or ink— how did it get into the egg or sperm? How did the egg or sperm "know" what the parent looked like? How could it encapsulate the parent's features? Darwin made this issue more complicated than he needed to since he believed—at least for a time—that off-spring could inherit features of their parents that their parents had acquired during their own lifetime. This notion—"inheritance of acquired characteristics"—had most famously been espoused in the early nineteenth century by Lamarck. The inheritance of acquired characteristics is now known to be profoundly untrue (at least for most purposes, although it still has a few champions). It now seems clear that the gametes develop entirely separately from the rest of the body and are not directly influenced by the body's strivings. Your father may have trained to be a brilliant flutist, but if you want to play the flute like him you must do your own practice. Darwin thought you could inherit your parents' acquired characteristics, however, and invented a mechanism that might make it work, which he called pangenesis. He suggested that every region of the body donated some fraction of itself, a kind of bulletin, to the developing gametes. This idea is actually close to what Hippocrates of Cos—he of the Hippocratic oath—had suggested at the turn of the fifth century B.C. Darwin's friends T. H. Huxley and John Lubbock warned him against this nonsense, but he seems to have persisted with it anyway.

Finally, the waters of the mid–nineteenth century were stirred by various forms of metaphysics. *Naturphilosophie* had diverse kinds of influence. Vitalism was powerful, the notion that the mechanisms of life required extra and special forces, over and above mere chemistry and physics. The most powerful metaphysical pressures came, however, from the Church.

THE CHURCH

Some of the clashes between the Church and the emerging science of the eighteenth and nineteenth centuries had a momentous quality: the general conflict between the concept of unwavering physical law and the biblical reports of miracles; the specific contradiction between the new geology and the creation story of Genesis. But there were other irritations, too, that may now seem rather comical, though they didn't at the time. In the eighteenth century the Swedish botanist Carolus Linnaeus, who established the "binomial" system for naming different creatures (as in *Homo sapiens,* the human being, and *Bellis perennis,* the daisy) and laid the foundations of modern classification, also demonstrated that flowering plants are sexual beings. The male element is the pollen, and the female lies within the ovary. No one doubts this now, and breeders and farmers of the time clearly knew this was the case and acted accordingly.

But some clerics did object to the idea. Jesus would surely not have singled out the lilies of the field for such praise if they had been practicing such covert naughtiness. Many biologists were clerics, too: you could not be a don unless you were ordained, and the cleric-naturalist had been a key figure in biology since the Middle Ages. Even by the middle of the nineteenth century the sexuality of flowering plants was not universally acknowledged. But for a breeder—and anyone seeking to understand heredity—such basics are a sine qua non.

So much for the unfolding science of heredity, in the decades and centuries before Mendel. The third thread in our story is that of the practical craftsmen-breeders. They contributed a great deal; and their practical knowledge—the ideas that they seemed to take for granted—often seemed to run far ahead of the scientists and philosophers. 'Twas ever thus. Science and technology work extraordinarily well when they work together, but they do not always march in unison.

THE BREEDERS

Scholars traditionally suggest that agriculture began around 10,000 years ago with the Neolithic revolution. That at least is when the first clear signs of cultivation appear in the archaeological record: permanent settlements like Jericho in Jordan and Çatal Hüyük in Turkey; caches of grain that are obviously different from wild types. I have argued, however, in *Neanderthals, Bandits, and Farmers* (Weidenfeld and Nicolson, 1998), that people must have been controlling their environment to a significant degree long before they were full-time farmers. The artifacts of the Neolithic do not reflect the true beginnings of farming; they merely show farming practiced on a scale that is big enough to show up in the records. The real beginnings almost certainly date back at least 30,000 years. As people controlled their environment, they influenced the kinds of plants that grew around them; for example, they might protect certain trees that had particularly juicy fruits from other predators. By this kind of selection, then, people have undoubtedly been "breeding" plants informally for many thousands of years. I suppose we could say that the first farmers truly became "breeders" when they first began *consciously* to encourage the particular plants and animals that they favored. When that was is anybody's guess, and always will be. I would not be at all surprised, however, if the first true breeders in this sense lived at least 30,000 years ago, and possibly a lot more.

Be that as it may, we know that by classical times, farmers were conscientiously breeding. Although we traditionally refer to "ancient" Romans, they were a modern people—a mere 2,000 or so years ago—and they were astute breeders, and were by no means the first. They distinguished between breeds of animals and varieties of plants and carried out "progeny testing": that is, they waited to see whether a particular bull or stallion was worth keeping for stud by assessing the quality of his first offspring. If his offspring were good, they kept him. If not, he was destined for the pot. Not every fine quality of an animal proves to be "heritable." Sometimes great-looking sires produce inferior offspring.

The crafts of breeding—selecting, progeny testing, and judicious crossing—were practiced on all fronts for many centuries: in serious agriculture, both crops and livestock; in horticulture; and in all kinds of hobbies in many countries—pigeons, fancy carp, and so on. For centuries, breeding remained a craft. But by the eighteenth century true science was beginning to creep in, and a key player in this was the naturalist J. G. Kolreuter (1733–1806).

Kolreuter was the first to carry out experiments in hybridization purely as an exercise in biology. His prime aim was to find out how organisms work rather than to make better crops and livestock, and in particular, he wanted to pin down the differences between "varieties" and "species." So he undertook a prodigious program of hybridization involving 54 species in 13 genera, which he published in a three-part report between 1761 and 1766, almost exactly a century before Mendel undertook his superficially less ambitious, but in the end more penetrating studies. Kolreuter believed, more or less as Aristotle did, that both parents produced a uniform, fluid semen that blended in the progeny, so that the offspring were intermediate between the two. These two kinds of semen had been "designed by the Creator for joining." When flowers formed, he suggested, the male and female semen separated out again. His studies certainly supported the eighteenth-century suggestion that plants are sexual beings.

Much more intriguingly, though, Kolreuter also showed that different characters *segregated* in the hybrid progeny. That is, a parent plant with white flowers and green pods crossed with a parent with yellow flowers and gray pods could give rise to offspring with white flowers and gray pods, or with yellow flowers and green pods. In other words, the different characters were inherited independently of each other. Mendel showed this too, a century later: it is recorded in the textbooks as one of his principal findings. Kolreuter also found, however, as all hybridists do, that the offspring often seemed to resemble one parent far more than the other, and this he put down to irregular mixing of semen. It is one thing to suppose that both parents contribute to the offspring and another entirely to infer that they both contribute equally on all occasions. Mendel cites Kolreuter in his paper of 1866.

Among the craftsmen-breeders of the eighteenth and early nineteenth centuries, three names are outstanding. In America, Joseph Cooper (1757–1840) was a remarkable breeder of crops, while in England, Robert Bakewell (1725–95) bred livestock, including the famous Dishley sheep for meat. But of most direct significance for Mendel was the Englishman Thomas Andrew Knight (1759–1838). Knight also—following Bakewell—bred cattle and sheep. Most important, though, were his experiments with cultivated plants.

From 1787, Knight sought to increase the productivity of crops by systematic hybridization, by crossbreeding, or "crossing," different types. Among other things, he emphasized the technique of artificial pollination. Furthermore, Knight recorded that of all the plants he hybridized, "none appeared so well calculated to answer my purpose as the common pea." The reason, he said, was that peas come in many forms, sizes, and colors, which makes them ideal for hybridizing, because hybridizing means beginning with parents with different qualities. But the pea was also ideal "because the structure of its blossom, by preventing the ingress of adventitious farina, has rendered its varieties remarkably permanent." (*Farina,* of course, literally means "flour" and was an early term for "pollen," used well before the true significance of pollen was appreciated.) Again, Mendel makes exactly this point in his paper of 1866. In modern parlance, we can say that garden peas are "inbreeders" (meaning they fertilize themselves) and so they are also "true breeding" (which means that unless they are officiously hybridized—cross-bred—the offspring closely resemble the parents). In Knight's work, then, we see the essential framework of Mendel's researches: hybridization of different types by artificial pollination, and, specifically, the use of peas, whose breeding can be so tightly controlled because the structure of their flowers prevents pollination from outsiders.

There is much more. Knight perceived that the traits of the parents were passed individually to the offspring. Mendel showed this too. Knight also observed—as, again, Mendel was to do— that when two uniform parents were crossbred, the first generation (F1) was uniform; but when the first generation was bred again (the different F1s crossed with each other), some traits that were present in the initial parent generation reappeared. He did not quantify his results in the way that Mendel did, but he observed the basic fact nonetheless.

Knight published his most influential paper on plant hybridization in 1799, and there he also states that hybrids between different species are generally infertile. Knight's paper was translated into German in 1800 and was well known on the Continent. There was certainly a copy in the university library at Brno and it bears the stamp of the Brno Agricultural Society. Did Mendel read it? It would be very surprising indeed if he did not.

But Mendel was not the first to pick up on Knight's work. Others included the Englishmen J. Goss and A. Seton, who also hybridized peas. In their publications of 1824 they demonstrated the phenomenon of dominance: how some traits in one parent completely obliterate other traits in the other parent. For example, *all* the progeny of a purebred yellow-flowered parent and a white-flowered parent could have yellow flowers. Mendel showed precisely this effect in his studies in the 1850s and 1860s. Goss and Seton then worked on other species, and so did Knight, who learned from them in this and followed their example. Mendel followed the same pattern: first peas, then other plants.

In Moravia, Mendel's country, breeders focused both on crops (such as apples) and sheep; again, they explored and developed the basic techniques and discovered the difficulties and limitations. Bakewell's work was obviously highly relevant to them, although his great creation—the Dishley—had been bred for meat, while the farmers of mainland Europe were more interested in wool. Since the seventeenth century, the wool farmers had focused on Spanish sheep: the merino. Books were published in the early eighteenth century on how to breed imported sheep. Crossing was recommended, and also consanguineous mating, although the Church, all-pervasive, objected to this on ethical grounds. Ferdinant Geisslern (1751–1824) was known as the Moravian Bakewell.

In the tradition of Kolreuter, breeders became more and more aware that they needed to supplement their empirical studies with science. Christian Andre wrote a zoology textbook in 1795 (in which he rejected preformation and asserted that both parents were involved in shaping the offspring). More grandly, in 1815, he drew up a general program of scientific development. He argued that we need to understand the basic mechanisms of the world if we are to exercise some control over our own affairs, citing Copernicus and Newton, who had brought science to the ancient craft of astronomy. If Moravians applied the same principles to their own concerns, said Andre, then all civilization would be indebted to them. His program echoes the sentiments of Francis Bacon, who said the same kind of things 200 years earlier.

As a good Moravian, Andre became involved in breeding sheep, which prompted him to think about heredity in general. In 1814, the year before he published his grand plan, he and H. F. Salm founded Moravia's Sheep Breeders' Society, which held meetings for breeders from throughout central Europe and included contributions on artificial selection and on the transmission of traits from parents to offspring.

Andre became involved in fruit, too, and in 1816 he drew attention to the work of Knight in breeding new varieties of vines and fruit trees. Out of these deliberations came Brno's Pomological and Oenological Association. Soon the association established a nursery to create new varieties, and Andre made contact with the Horticultural Society of London, of which Knight was president, and with the Pomological Association at Altenberg (near Leipzig), where G.C.L. Hempel was secretary.

Hempel, too, was forward-looking. In 1820 he wrote that "higher scientific pomology" was moving toward the point at which breeders might create new varieties at will, according to specifications for size of fruit, shape, color, and flavor. This dream is now coming true in precisely the way Hempel envisaged. First, though, he said, it would be necessary to understand the rules of heredity, what he called the laws of hybridization. For this, "a new type of natural scientist" would need to emerge: "a researcher with a profound knowledge of botany and sharply defined powers of observation who might, with untiring and stubborn patience, grasp the subtleties of these experiments, take a firm command of them, and provide a clear explanation." Hempel might have been describing Mendel, who in fact was born two years after he made this statement. Truly, we can add Knight (at a distance), Andre, and Hempel to the pantheon of Mendel's inspirers. Taken together with the great biologists already cited, his was almost an embarrassment of riches.

Indeed, the world at large caught the bug of agricultural science in the late eighteenth and early nineteenth centuries. It appeared as a subject on university curricula. Edinburgh was among the first to introduce it and has remained in the forefront, giving rise to Roslin Institute, which, in our own day, produced Dolly, the cloned sheep. In Moravia, the ever active Andre suggested setting up a professorship for agriculture in 1808. The first was established in 1811 at Olomouc, where Mendel attended the Philosophy Institute, and the second in 1816 at Brno. In 1823, J. K. Nestler (1783–1841), a former colleague of Andre's, began teaching breeding at Olomouc, and in 1824, F. Diebl, friend and associate of Abbot Napp, took the chair at Brno. Principally a plant scientist, Diebl published a five-volume book on agricultural science between 1835 and 1844 in which he described artificial pollination. With Nestler, he coauthored *General Natural History* in 1836, and both of them published articles on hybridization and heredity. In the 1830s, Diebl worked with Abbot Napp on the committee of the Agricultural Society and especially with the Pomological Association, when Napp was president. Diebl too, then, was a significant and influential figure.

Nestler, for his part, was interested mainly in sheep. In his writings he referred to Kolreuter and discussed progeny testing and consanguineous breeding. In his lectures he spoke both of blending inheritance (where characters combine features of both parents) and of characters that have an either-or quality. Thus cattle tend to either have horns or not—not somewhere in between. He also referred to "sports," which he thought were inexplicable. Whether offspring could inherit features that their parents had acquired in their own lifetime, as Lamarck had proposed, was, he said, "problematical." In the mid-1830s he told Brno's Sheep Breeders' Association that "the most essential thing of all for improved sheep breeding as well as being an urgent question of our time [is] the ability to inherit."

The science of heredity was undoubtedly advanced by the breeding of sheep. In 1812, Christian Andre described artificial selection in sheep to increase wool production, and in 1816 his son, Rudolf, published a manual on sheep breeding. In 1818, Andre senior perceived that consanguineous mating would weaken the breed, which, he said, was "a natural physiological law," and in that same year he asked a renowned Hungarian sheep breeder, Count E. Festetics (1764–1847), to formulate the main principles of interbreeding in breeding practice. The following year, Festetics published what he called genetic laws—*genetische Gesetze.* The term *genetic* is generally supposed to be a twentieth-century invention, but this is evidently not the case. Festetics's term *genetische* is the same word.

Festetics's "laws" are not really laws at all: they are observations. But they are sensible observations that definitely helped consolidate the subject. First, he observed that the characteristic traits of healthy parents are indeed inherited by the progeny. Second, however, the traits of one of the grandparents may also appear in the progeny, even though they may not have been present in the parents. But then again, the progeny may exhibit traits that are quite different from either parent; and if these traits do not correspond to the aims of the breeder, but are heritable, then of course they are undesirable. Finally, if you are going to inbreed—indulge in consanguineous matings—then you must, said Festetics, select the stock animals scrupulously. In general, consanguineous mating was necessary to maintain constancy; but it had to be used circumspectly if production was to be maintained. He also observed, as early as 1820, that wool quality was now judged "with mathematical precision."

All in all, there has been no better place and time to discuss the breeding of sheep and all the ideas that may follow from it than Moravia in the 1830s. Nestler boasted that "Moravia can claim special credit for having become . . . a source of modern, rational sheep breeding." After that decade, however, Moravia's sheep industry waned somewhat, being outcompeted by cheap imports from the British colonies. Nestler himself died suddenly in 1841.

For Abbot Napp, fruit was his first love. In 1840 he chaired the fourth congress of German-speaking farmers and foresters in Brno and there discussed hybridization as a way of creating new varieties, though others said that crossbreeding merely produced random results. A year later Napp commented: "Nothing certain can be said as to why production through artificial fertilisation remains a lengthy, troublesome, and random affair." Breeding and the mechanisms of heredity were the burning issues of mid-nineteenth-century Moravia, but despite the many advances—all the straws in the wind that can be seen in retrospect—puzzlement reigned.

In the early nineteenth century, too, various learned bodies offered prizes to anyone who could solve problems of heredity and breeding. Many of these were clearly grist to Mendel's mill. In 1822 the Berlin Academy of Sciences offered an award to anyone who could show for all time whether plants are, or are not, sexual beings. A. F. Wiegmann, an apothecary from Brunswick, was the winner in 1828. He crossed various species of peas and produced some hybrids that more closely resembled one parent or the other, some that combined traits of both, and some that seemed to bear no resemblance to either parent. He seems thereby to have answered the main question; plants must be sexual, since both may contribute to the offspring; again, though, his studies did not show definitively whether both parents always contribute equally. His experiments have much to commend them, however. Note again, though, his choice of peas.

A similar prize was offered in 1830 by the Haarlem Academy of Sciences, this time won by F. C. Gartner (1772–1850). In the 1830s and 1840s, Gartner carried out more than 10,000 artificial fertilizations in 700 species of plant, yielding 250 different hybrids, and he summarized his findings in an extensive monograph in 1849. Gartner cited both Kolreuter and Knight. Like them, he showed that hybrids between different species tend not to be fertile, and in general he reinforced the idea that species are indeed constant and cannot be changed into other species. Mendel bought Gartner's monograph in the year that he wrote it, and again, he quoted him in his paper of 1866.

Purkinje's friend and correspondent R. Wagner (1805–64) at the University of Göttingen also deserves mention. In 1853 he argued that the rules of heredity might be discerned by analyzing a mass of data statistically, although he felt this would be expensive and time consuming. He was right; but this, of course, is precisely what Mendel undertook. Wagner thought that both parents contribute to heredity, an idea he got from Purkinje.

Finally, after 1850, both Franz Unger at the University of Vienna and Carl Naegeli at the University of Munich studied plant hybridization from a physiological point of view, both trying to make use of physics and chemistry. Unger was Mendel's teacher and a great admirer of Naegeli, and Mendel tried to interest Naegeli in his own research, though with miserable results. If only Naegeli had grasped the significance of Mendel's work, the course of science would have been quite different. Genetics would have begun 40 years earlier.

By the time Mendel began his own studies into heredity some time in the mid-1850s—probably in 1854—a huge amount of groundwork had already been done. There were so many fine ideas around that it may seem as if he had merely to dot the i's and cross the t's. The crafts of breeders were well advanced, but many were stressing the need for keen experimentation, and indeed, such experiments had begun with Kolreuter in the eighteenth century. Specifically, these early experimenters stressed the need to explore heredity by hybridization: crossing plants (or, in principle, animals) that were sexually compatible but had somewhat different characters. They also emphasized the need to work with true-breeding plants—those whose offspring really did resemble them when they were self-fertilized—and the garden pea had long since emerged as the favored subject. Indeed, the garden pea occupied the same niche in early-nineteenth-century breeding experiments as the fruit fly, *Drosophila,* assumed in the twentieth century. Finally, Wagner, in particular, emphasized the need for large-scale experiments and statistical analysis.

Mendel's education, predilections, and working conditions primed him beautifully for the task. He was born into a community of gardeners and plant breeders and was surrounded by them all his life. But he was a physicist and a passable mathematician— used to the idea that certainty in science depended on quantification. I feel, too (though this is only a feeling), that a physicist would take naturally to the idea that heredity is particulate—digital—in nature, that creatures would pass units of heredity from generation to generation. Biologists are more inclined to think in terms of the mixing of inks, semen, or "sanguineous nutriments." Mendel also, clearly, had the scientist's instinct: to carry out simple experiments that would give clear answers and then work outward from areas of certainty into more difficult problems.

Darwin, the greatest biologist of the nineteenth century, (and indeed of all time) failed to see the underlying simplicity of heredity partly because he had the temperament of a naturalist—focusing on and relishing the *variety* of nature. Darwin, in fact, as he explains in *Origin of Species,* spent a lot of time talking to breeders of pigeons. But it is impossible to see the patterns of heredity in such animals, for all kinds of reasons (unless you know in advance what patterns you should be looking for). As we will see in the next chapter, Mendel was able to see simple hereditary ratios in his peas partly because the peas were true breeding (he went to great lengths to ensure that this was so) and partly because they produce enormous numbers of offspring. The simple ratios are of a statistical nature: they are evident only when the numbers *are* enormous. When creatures produce too few offspring, as pigeons or human beings do, and when they are not true breeding themselves, the simple patterns are not evident at all.

But although in retrospect we can see that Mendel picked up on an enormous history of good, solid ideas, confusion reigned nonetheless. To be sure, a great deal of good and necessary work had been done. But the good and necessary results had not been brought together to create a coherent picture; and the good ideas were balanced, almost equally, by traditional notions that were crackpot or just plain wrong. It was still not 100 percent clear—at least not to everybody—even that plants were sexual; or if they were, whether each parent (plant or animal) always contributed in equal amounts. These are very basic matters to be uncertain about.

Darwin summarized the prevailing state of confusion in *Origin of Species* in 1859—when Mendel's experiments were well advanced but not yet published. He wrote:

> . . . no-one can say why the same peculiarity in different individuals . . . is sometimes inherited and sometimes not so: why the child often reverts in certain characters to its grandfather, or other much more remote ancestor; why a peculiarity is often transmitted from one sex to both sexes, or to one sex alone, more commonly but not exclusively to the like sex.

Even in 1872 the renowned animal breeder H. Nathusius still felt constrained to observe that:

> . . . the laws of heredity have not yet fallen from the tree of knowledge—that which, as legend would have it, led Newton on the right path to the conception of the Law of Gravity.

In truth, the essence of the laws *was* known then, at least to Mendel, and he had made the laws known in lectures and publications. But nobody was paying attention.

Others did make various kinds of progress between 1866, the year of Mendel's paper, and 1900, the year of its rediscovery. Chromosomes were described in that period, and in the 1880s the details of cell division, which involves the orderly separation of chromosomes, were first outlined. In 1866 yet another great German, August Weismann (1834–1914), declared that the germ plasm (the tissue that gives rise to the gametes) develops entirely separately from the rest of the body, so that changes within the body cells could not be passed on to the next generation. This assertion effectively showed that Lamarck's notion of the inheritance of acquired characteristics was impossible. At least it is impossible in animals: some plants may produce new flowers, and hence eggs and pollen, from just about any tissue. In 1883, W. K. Brooks summarized the general state of understanding in the second edition of his *Law of Heredity*. Brooks was one of Thomas Hunt Morgan's early instructors. Francis Galton—also working with peas—produced his "ancestral law of heredity" which he published in 1898. This was well received, although as R. A. Fisher showed 20 years later, Galton's "ancestral law" merely applied Mendel's principles to characters that are coded by more than one gene (that is, to "polygenic characters," as discussed in the next chapter). More broadly, Galton developed the discipline of "biometrics," which discussed continuous variations between different individuals—in essence, the very characters that Mendel avoided during his studies on peas because of the intractable complexities he knew would ensue. However, Galton's biometrics, carried forward in particular by his disciple Karl Pearson, has formed a parallel tradition of heredity throughout the twentieth century; it gave rise, among other things, to the eugenics movement. As this book specifically traces the Mendelian influence, the science of biometrics will be discussed only briefly, in chapter 9.

FOUR

From Mendel to Molecules

Mendel is pivotal to genetics in the way that Charles Darwin is pivotal to biology as a whole. Everything before Mendel seems a little primitive—often excellent, and much of it crucial, but mainly now of historical interest. The modern age begins with Mendel. Put his ideas and Darwin's together, as was achieved in the early twentieth century, then stir in a little molecular biology (which is the great contribution of the later twentieth century), and we have the core of modern biology: what relativity and quantum mechanics are to physics. Since Mendel's and Darwin's ideas are not in principle difficult, biology emerges as an easy subject. It is also vital. It's all very pleasing.

Mendel conducted his most important experiments—on the garden pea, *Pisum sativum*—in the monastery garden at Brno between around 1856 and 1864, although he seems to have carried out two years' preliminary trials, in 1854 and 1855, to ensure that the peas he was using were true breeding, or to anticipate a term explained later, "homozygous." He reported his findings to the Brno Natural Science Society in February 1865 and again in March 1865, and they were published formally as a paper in the journal of the society in the following year, 1866, with the title "Versuche über Pflanzen-Hybriden"—that is, "Experiments in Plant Hybridization." This is a stunning paper, lucid and thorough, and must have been seen by the greatest biologists in Europe, most of whom were attached to the Brno society. Yet it was all but ignored. Mendel's disappointment must have exceeded even that of David Hume, who mourned a century earlier that his maiden masterwork, *A Treatise of Human Nature,* had fallen "deadborn from the press." Mendel's work was rediscovered in 1900, but by that time he had been in his grave for 16 years.

After his death, insult was added to the disappointment of his last years. Political insults were heaped on his head in eastern Europe, but even here in the West, where his work has found its home, many have suggested that his research was simply lucky; in fact, it has almost become orthodox to argue this. If he had not chosen to study peas, say the critics, he would not have achieved such clear-cut results. Given that he did study peas, he still would not have produced such sharp results if he had not singled out the particular characters (physical features) that he happened upon: flower color, seed color, and so on. Other factors, such as linkage and polygenic characters, would have confused the picture no end. As a coup de grâce, the great British statistician-biologist R. A. Fisher suggested in 1936 that Mendel's results were just too good to be true, that ". . . most, if not all, of the experiments have been falsified so as to agree closely with Mendel's expectations." Fisher fully acknowledged Mendel's genius—perhaps it really does take one to recognize another—and never for a second doubted Mendel's probity. He did not suggest that Mendel himself had done the falsifying. But others, less scrupulous and far less scholarly, have been all too keen to suggest that Mendel cheated.

However, Mendel's experiments with peas were not "lucky" at all. He was a fine biologist and had many excellent predecessors, and he knew exactly what he was doing. As his paper clearly shows (have none of his critics actually read it?) he *did* confront linkage *and* polygenic characters (in beans) and dealt with both effortlessly. For my part, I am again inclined to quote David Hume, who said of people who report miracles: "If the falsehood of his testimony would be more miraculous than the event which he relates then, and not till then, can he pretend to command my belief or opinion." In other words, the idea that Mendel cheated is more incredible than any alternative explanation, even if we don't know for certain what the alternative explanation might be.

Initially, however, it is important to understand what Mendel actually did with his peas, and to point out the significant features that often seem to be overlooked. Mendel knew that he could not simply assume that the rules of heredity he found in peas would *necessarily* apply to all other plants—or indeed to creatures apart from plants. As he says in his paper of 1866: ". . . a generally applicable law governing the formation and development of hybrids . . . can only be arrived at when we shall have before us the results of detailed experiments made on plants belonging to the most diverse orders." So from the 1860s onward, he set out to work through plants of other families—about 17 species in all—and in his later life his work with bees showed him to be both a master technician and an extraordinarily accomplished naturalist. But he never finished any of this later work, apart from what he called some minor experiments with *Phaseolus* (kidney) beans, which he carried out at the same time as the work on peas. As he said in his paper, "It requires some courage to undertake a labour of such far-reaching extent," and he was talking only about his work on peas. In truth, his grand agenda—to find the basic patterns of heredity in all species—would have required at least a lifetime, and a team of senior scientists, postgrads, and graduate students, such as a modern Genetics department might bring to bear.

Mendel had the peace that many modern scientists crave but not the resources.

In the second half of this chapter, it becomes clear how all the modern ideas of twentieth-century classical genetics flow naturally from Mendel's initial observations. There were many twentieth-century greats, among whom we can single out Thomas Hunt Morgan (1866–1945), who at Columbia University in 1908 began the research on the fruit fly *Drosophila* that continues to this day, and his onetime colleague, the irascible Hermann Joseph Muller (1890–1967), who greatly elucidated the mechanisms and importance of genetic mutation, the process by which genes actually change (and then either do something different or simply become nonfunctional). But as the now-standard ideas of classical genetics have been well described elsewhere, I want to move as quickly as possible to the issues that mainly concern us now.

MENDEL'S PEAS

Mendel's key experiments were an exercise in hybridizing garden peas: crossing different varieties that had clearly distinct physical characters. There was no "luck" in his choice, either of experimental design or of subject. Many biologists before him had stressed that the problems of heredity would be solved only by hybridization of different, clearly defined varieties; and some—Kolreuter, Seton, Goss, and others—specifically recommended peas.

Mendel began his preliminary work in 1854 ". . . with 34 more or less distinct varieties of Peas . . . obtained from several seedsmen." These were then ". . . subjected to two years' trial." Then, "for fertilisation, 22 of these were selected and cultivated during the whole period of the experiments." Finally, as he announced in 1865, ". . . after eight years' pursuit [the work is] concluded in all essentials." Thus the experiments themselves ran from 1856 to 1864.

Mendel makes it very clear that not all plants would serve his purposes: For, he said:

The experimental plants must necessarily:
1. Possess constant differentiating characteristics.
2. The hybrids of such plants must, during the flowering season, be protected from the influence of all foreign pollen, or be easily capable of such protection.

He adds as an afterthought, although he might have made it point 3:

The hybrids and their offspring should suffer no marked disturbance in their fertility in the successive generations.

So why home in on peas? Well, says Mendel, while acknowledging Kolreuter, "At the very outset special attention was devoted to the Leguminosae. Experiments which were made with several members of this family led to the result that the genus *Pisum* was found to possess the necessary qualifications."

Let's look at these qualifications one by one: a pedantic exercise, maybe, but the ghost of luck should be laid forever. On the first point: as Mendel and about a million gardeners were perfectly well aware, many if not most common plants are *not* particularly "constant" in their "characters." At least, some are too constant: all primroses are yellow, for example, and if all the plants have the same color, then it is impossible to see any patterns of heredity, since there are no variations to analyze. Others, though—lupins, snapdragons—have very variable colors, and a parent of any one color seems to give rise to offspring of many different colors. When the characters are *so* variable, any underlying pattern is obscured by complexity, and indeed by randomness. So to make sense of heredity, the experimenter needs plants that come in several or many different forms, but each of those forms must breed true, unless specifically hybridized. The interaction of round-seeded plants with wrinkled-seeded plants can be assessed by crossing them, but *only* if the round-seeded plants normally give rise to round-seeded offspring, and the wrinkled-seeded plants to wrinkled-seeded offspring. If round-seeded plants produce wrinkled-seeded offspring when self-fertilized, or wrinkle-seeded plants produce round-seeded offspring, then next to nothing can be learned by hybridizing the two types. If either type, when "selfed," could also produce offspring of quite different types, then it would be impossible to make any sense of the outcome. Peas fit this particular bill, but remarkably few other plants do. Mendel knew this. But he needed to carry out two years' preliminary work to ensure that the different samples of peas he began with were indeed true breeding, that plants of any particular type, when left to themselves, did indeed produce offspring only of their own kind.

On the second point: "Accidental impregnation by foreign pollen . . . would lead to entirely erroneous conclusions." With peas and many other legumes, says Mendel, it is particularly easy to avoid this: "On account of their peculiar floral structure . . . a disturbance through foreign pollen cannot easily occur, since the fertilising organs are closely packed inside the keel and the anthers burst within the bud, so that the stigma becomes covered with pollen even before the flower opens. This circumstance is especially important." In fact, Mendel showed that unless his peas were attacked by a particularly pestilential beetle called *Bruchus pisi,* which caused the flower to burst open, they were virtually never pollinated by other peas that just happened to be in the neighborhood.

All the fertilizations were carried out either by "selfing" (leaving the flowers alone and allowing the pollen to scatter onto the stigmas in the same flower) or by deliberate hybridization using artificial pollination, the technique that had been in wide use since around 1820. Of this Mendel says, "Artificial fertilisation is certainly a somewhat elaborate process, but nearly always succeeds. For this purpose the bud is opened before it is perfectly developed, the keel is removed, and each stamen carefully extracted by means of forceps, after which the stigma can at once be dusted over with foreign pollen."

Finally, on his third point, Mendel says, "Reduced fertility or entire sterility of certain forms, such as occurs in the offspring of many hybrids, would render the experiments very difficult or entirely frustrate them." Putting the matter crudely, it would be very difficult to produce worthwhile statistics on patterns of inheritance if a proportion of the experimental plants refused to breed at all, or did so only in a fitful way that defied analysis.

Once he had identified "lines," as a modern breeder would say, of peas that he knew were true breeding, he very deliberately selected characters for study that he knew would give comprehensible results. There was no luck in this. He was simply aware of the principle that Sir Peter Medawar made explicit a century later: "Science is the art of the soluble." Again, Mendel clearly describes why he chose the features he did:

> The various forms of Peas selected for crossing showed differences in length and colour of the stem; in the size and form of the leaves; in the position, colour, size of the flowers; in the length of the flower stalk; in the colour, form, and size of the pods; in the form and size of the seeds; and in the colour of the seed-coats and of the albumen [by which he means the seed leaves, or cotyledons]. Some of the characters noted do not permit of a sharp and certain separation, since the difference is of a "more or less" nature, which is often difficult to define. Such characters could not be utilised for the separate experiments; these could only be applied to characters which stand out clearly and definitely in the plants.

So which particular characters did he choose to study? Again, in his own words (though the passages in square brackets are mine):

The characters which were selected for experiment relate:

1. To the *difference in the form of the ripe seeds.* [They were either round or wrinkled.]

2. To the *difference in the colour of the seed albumen (endo-sperm).* [Either pale yellow, bright yellow and orange, or intense green.]

3. To the *difference in the colour of the seed-coat.* This is either white, with which character white flowers are constantly correlated; or it is grey, grey-brown, leather-brown, with or without violet spotting, in which case the colour of the standards is violet, that of the wings purple, and the stem in the axils of the leaves is of a reddish tint. The grey seed-coats become dark brown in boiling water.

In this instance Mendel is almost certainly observing the phenomenon of "linkage": the fact that sometimes characters are not inherited entirely independently of all other characters, but are often (or always) associated with other characters. As he says here, particular pod colors do tend to be associated with particular flower colors: for example, white pods go with white flowers. In truth, this might not be linkage—it might simply be an example of one gene producing two different effects. But Mendel did not achieve clear results just by striking a lucky course through all the difficulties. He perceived the difficulties and dealt with each one sensibly. To continue:

4. To the *difference in the form of the ripe pods.* [Either "inflated," or "deeply contracted in places."]

5. To the *difference in the colour of the unripe pods.* [Either light to dark green, or vivid yellow.]

6. To the *difference in the position of the flowers.* [Either axial, meaning along the main stem, in the angles of the leaves, or terminal, meaning at the end of the stem.]

7. To the *difference in the length of the stem.* [This was "very various in some forms." But it was constant enough when plants were grown in constant conditions. Mendel differentiated between very tall ones— 6 to 7 feet long—and markedly short ones—between 9 and 18 inches.]

This, then, was the basic setup: 22 varieties of well-selected peas that he knew were true breeding, and seven carefully selected characters to study that "stand out clearly and definitely in the plants." For the next eight seasons he studied various hybridizations between the different types. In all, he minutely studied about 10,000 different plants. He knew that the patterns he would be looking at were of a statistical nature, and unless he looked at a lot he would not achieve valid results. This was the mathematician in Mendel coming to the fore.

Because Mendel's experiments get more complicated as the generations pass, and he takes at least 10,000 words to describe them, it is not possible to examine all of them in detail. It is easier to look at just a few, to illustrate the main points. I have decided to cheat a little here and there by employing twentieth-century vocabulary to make it easier to see what is going on. In particular, Mendel makes clear his realization that characters are passed on from generation to generation in the form not of different inks (as Darwin once supposed) but of discrete "factors," sometimes known as "Mendelian factors." We now call these factors genes. But although the term *gene* was not coined until 1909 (by Wilhelm Johannsen), I will use it where appropriate in the following account.

To begin with, Mendel simply crossed plants with each kind of character with other plants that had the corresponding but different character: round-seeded with wrinkled-seeded; yellow cotyledons with green cotyledons; white seed coat with gray seed coat; inflated pods with contracted pods; green pods with yellow pods;

axial flowers with terminal flowers; long stem with short stem. With each cross he found that *all* the offspring had the character that is cited first in the above list. Thus when true-breeding round-seeded plants were crossed with wrinkled-seeded plants, *all* the offspring had round seeds, and so on.

From this, Mendel derived one of his most significant principles, which has resonated through all genetics ever since:

> Henceforth in this paper those characters which are transmitted entire, or almost unchanged in the hybridisation, and therefore in themselves constitute the characters of the hybrid, are termed the *dominant,* and those which become latent in the process *recessive.* The expression "recessive" has been chosen because the characters thereby designated withdraw or entirely disappear in the hybrids, but nevertheless reappear unchanged in their progeny.

The idea of "dominant" and "recessive" genes is now a familiar one. Here, in Mendel's paper of 1866, is the first clear mention. Specifically, he says,

> Of the following characters which were used in the experiments the following are dominant:
> 1. The round or roundish form of the seed.
> 2. The yellow colouring of the seed albumen [cotyledons].
> 3. The grey, grey-brown, or leather-brown colour of the seed-coat, in association with violet-red blossoms and reddish spots in the leaf axils.
> 4. The simply inflated form of the pod.
> 5. The green colouring of the unripe pod.
> 6. The distribution of the flowers along the stem.
> 7. The greater length of the stem.

On this last point, Mendel observed the phenomenon we noted in the last chapter: "hybrid vigor." For the purebred tall plants were never more than 7 feet tall, while the hybrid ones—those crossed with short-stemmed plants—were up to 7½ feet tall. *Sometimes* when two individuals with contrasting characters are crossed, the characters of the offspring average those of the parents. But sometimes, emphatically, they do not. In this case—because of hybrid vigor—the offspring are *taller* than the taller parent, even though the other parent is extremely short!

In general, Mendel says, "Experiments which in previous years were made with ornamental plants have already afforded evidence that the hybrids, as a rule, are not exactly intermediate between the parental species. With some of the more striking characters, those for instance which relate to the form and size of the leaves, the pubescence of the several parts, etc., the intermediate indeed is nearly always to be seen. In other cases, however, one of the two parental characters is so preponderant that it is difficult or quite impossible to detect the other in the hybrid. This is precisely the case with the Pea hybrids."

Now—perhaps remembering his bad experience with Fenzl, when he failed his teaching exams for the last time—a note of triumph seems to creep into his paper.

It was furthermore shown by the whole of the experiments that it is perfectly immaterial whether the dominant character belongs to the seed plant or to the pollen plant; the form of the hybrid remains identical in both cases. This interesting fact was also emphasised by Gartner.

Fenzl was a latter-day spermist, clinging to the idea that all the inheritance in plants comes from the pollen, the male parent. Mendel was delighted to demonstrate clearly what he knew all along: that this simply is not the case. The contributions of male and female are equivalent (except of course in beastly organisms like *Hieracium* that practice parthenogenesis, as noted in chapter 2). Actually, there are caveats here. In particular, research from the 1980s onward has shown that in some cases, at least in mammals, the father or the mother puts his or her own stamp on particular genes. But the usual, default position is entirely as Mendel says: it really doesn't matter which parent contributes which genes.

Mendel now allowed the hybrid offspring to self-fertilize, as peas do when they are left alone. He then observed and counted the resulting offspring in each case. Again, in his own words:

Expt 1. Form of seed: From 253 hybrids 7324 seeds were obtained in the second trial year. Among them were 5474 round or roundish ones and 1850 angular wrinkled ones. Therefrom the ratio 2.96:1 is deduced.

Expt 2. Colour of albumen: 258 plants yielded 8023 seeds, 6022 yellow, and 2001 green; their ratio, therefore, is as 3.01:1.

So he worked through all seven pairs of characters—which led to the grand conclusion:

If now the results of the whole of the experiments be brought together, there is found, as between the number of forms with the dominant and recessive characters, an average ratio of 2.98:1, or 3:1.

This three-to-one ratio is another crucial Mendelian finding. Mendel describes how he thought it was arrived at, and as always, it seems, his instincts and reasoning were bang on the button.

Mendel explains the reason for this ratio* but it seems reasonable to summarize the argument in a more modern form. Herein lies the crux of classical genetics.

For imagine, now, that each *simple* character—characters such as round-seededness in peas—is conferred by a single factor (which we now call a gene). Now we have merely to surmise that the complete set of all characters in any creature is conferred by its genes. Now imagine that each individual contains *two* complete sets of the genes appropriate to its species. One set has been inherited from one parent, and one set from the other. Now imagine what happens when the individual reproduces. It produces gametes: sperm or eggs. We merely have to suppose that each gamete contains just one set of genes. Then, after the genes in the sperm combine with the genes in the egg during conception, the resulting embryo again contains two complete sets of genes: one from each parent.

So now let's look in genetic terms at what happens when a round-seeded pea is crossed with a wrinkled-seeded pea. I will use my own nomenclature—"R" for the gene for round-seededness, and "Wr" for the gene for wrinkled-seededness—simply because I think it makes things easier.

A purebred true-breeding round-seeded pea contains two copies of the R gene: we can call them RR. A purebred true-breeding wrinkled-seeded pea contains two copies of the Wr gene: we can call them WrWr. So far so easy.

---

*In fact, Mendel does describe gametes, with different characters, though the sentence in which he does so lacks some of his usual lucidity. Perhaps it works better in German. Anyway: "Since the various constant forms are produced in *one* plant, or even in *one* flower of a plant, the conclusion appears logical that in the ovaries of the hybrids there are formed as many sorts of egg cells, and in the anthers as many sorts of pollen cells, as there are possible constant combination forms, and that these egg and pollen cells agree in their internal compositions with those of the separate forms."

Note, though, how confidently he talks about egg cells and pollen cells. In Mendel's day such ideas were new. As noted in the last chapter, the fusion of egg cells with sperm cells was observed in plants only in the 1850s, and not in flowering plants but in algae.

All the gametes—pollen or eggs—produced by the true-breeding round-seeded pea contain the R gene; *all* the gametes produced by the wrinkled-seeded pea contain the Wr gene. Still very easy.

The hybrids produced by crossing true-breeding round-seeded plants with wrinkled-seeded plants will *all* contain one R gene and one Wr gene. These can be called RWr. All the hybrid, RWr individuals have round seeds because the R gene dominates the Wr gene.

Now we allow RWr plants to cross with other RWr plants, or rather, we allow them to self-fertilize, which amounts to the same thing.

Each (RWr) plant produces *two* kinds of gamete: some containing an R gene, and some containing a Wr gene. That is, some pollen and some eggs contain R genes; and some pollen and some eggs contain Wr genes.

There are now four possible combinations, each of which is equally likely to occur. An R pollen may combine with an R egg to produce RR offspring. An R pollen may combine with a Wr egg to produce RWr offspring. A Wr pollen may combine with an R egg to produce WrR offspring. Or a Wr pollen may combine with a Wr egg to produce WrWr offspring.

However, the RWr offspring are genetically the same as the WrR offspring because it does not matter which gene comes from which gamete. So now we have three genetically distinct kinds of offspring: RR, RWr (or WrR), and WrWr. Since each combination is equally likely to occur, we can see that for every four offspring— on average—we can expect one RR individual, two RWr (or WrR) individuals, and one WrWr individual. Thus these different genetic types occur in the ratio 1 to 2 to 1, generally written 1:2:1. This is another famous Mendelian ratio. Of course, we don't expect to see this ratio of 1:2:1 if we produce only four offspring. We will see such ratios only when we produce a great many offspring, which Mendel knew perfectly well and which is why he set out from the beginning to produce large numbers.

Finally, as we have also seen, the RWr's and the WrR's both have round seeds, because the R gene dominates the Wr gene. Hence the RWr's and the WrR's look exactly the same as the RR's. Hence the ratio of three to one that Mendel observed when he crossed round-seeded plants with wrinkled-seeded plants (or indeed, crossed any pair of plants with contrasting dominant and recessive characters). There is no way to tell that the RR's are genetically different from the RWr's or the WrR's until further breeding experiments are done. The offspring of self-fertilized RR's will all have round seeds; but *some* of the offspring of the self-fertilized RWr's or WrR's will have wrinkled seeds. Mendel's two-year preliminary experiments were conducted to ensure that all the round-seeded peas he began with were RR's rather than RWr's or WrR's. Already, albeit explained with some reference to modern terminology, we have most of Mendelian genetics. There is one further vital principle. In general, though not invariably in practice, different characters are inherited *independently* of each other. Mendel demonstrated and described this. Later geneticists called this Mendel's law of independent assortment (although Mendel himself did not formally present this "law" as such).

Among other combinations, Mendel crossed plants that had round seeds and yellow cotyledons with plants that had wrinkled seeds and green cotyledons. All the hybrids then contained genes for round seeds and yellow cotyledons *and* genes for wrinkled seeds and green cotyledons. He then allowed these hybrid offspring to self-fertilize. Then, as always, he counted the offspring. This is what he reports:

In all, 556 seeds were yielded by 15 plants, and of these there were:
315 round and yellow
101 wrinkled and yellow
108 round and green
32 wrinkled and green

If you examine these figures, you see that they give us another famous "Mendelian ratio": 9:3:3:1, or nine to three to three to one.

To describe in words why we obtain this ratio would take too many pages. You can probably see the reason intuitively, but as an aid to thought we can present the whole thing as a Latin square. We simply write all the possible gametes along each axis and see what the combinations look like. As we see, there are four possible kinds of gamete, giving us sixteen possible combinations. Nine out of the sixteen contain at least one gene for roundness and at least one for yellowness, so the seeds appear round and yellow. Three contain two genes for wrinkliness and at least one gene for yellowness, and so emerge wrinkled and yellow. Three contain at least one gene for roundness and two genes for greenness and so are round and green. Finally, a miserable one out of the whole sixteen contains two genes for wrinkliness and two for greenness, and so is wrinkled and green.

These figures are simply ratios. You might well produce sixteen offspring—or, though this is less likely, 160 offspring—and produce no wrinkled and green offspring at all. This is why some combinations of characters seem to disappear and reappear so capriciously in the family trees of creatures that do not have huge numbers of offspring. Human beings have few offspring, which is why—as Darwin noted—it is very difficult indeed to see the Mendelian ratios within any one family tree. Of course, you can see Mendelian patterns in human families if the genealogical tree is large enough and you know in advance what you are looking for, which is why genetic counseling works. You can figure out for yourself how complex the genetic patterns and ratios become if you consider combinations of three, or four, independent characters at once. You need a very large Latin square to do this. If you were to consider, say, five or six characters, you would need a two-year prison sentence with nothing else to occupy your time to do full justice to the possibilities. In short, Mendel's very simple rules readily produce immensely complicated outcomes. Mendel was well aware of this. His genius was to trace back from the surface complexity to the underlying simplicity and then work outward again.

Clearly, Mendel knew exactly what he was doing, and why, and states his reasons with exemplary clarity. Yet one last crumb of calumny remains to be answered: some critics say it was fortunate that Mendel apparently never encountered any characters of a polygenic nature—that is, the kind that are shaped by more than one gene. Such characters are generally not inherited in the all-or-nothing pattern on which Mendel based his rules.

But of course Mendel did encounter such characters. In his initial pea experiments, he describes them and then deliberately puts them to one side because he knew they would introduce unnecessary complications before he had acquired the theoretical equipment to deal with them. Such polygenic characters would have included at least some of those that he said were of a "'more or less' nature, which is often difficult to define." But in what he calls his "minor experiments" with two species of *Phaseolus* beans, he confronts this problem head-on. He found that when beans of different colors are crossed, the results are complicated:

> Apart from the fact that from the union of a white and a purple-red colouring a whole series of colours results, from purple to pale violet and white, the circumstance is a striking one that among 31 flowering plants only one received the recessive character of the white colour, while in *Pisum* this occurs on the average in every fourth plant.

A blow to the theory worked out in peas? Not at all; for with remarkable anticipation of twentieth-century ideas Mendel writes:

> Even these enigmatic results, however, might probably be explained by the law governing *Pisum* if we might assume

that the colour of the flowers and seeds of *Ph. multiflorus* is a combination of two or more entirely independent colours, which individually act like any other constant character in the plant.

What could be clearer? But of course, Mendel being Mendel, he knew that such easy explanations must be explored in greater depth:

It would be well worth while to follow up the development of colour in hybrids by similar experiments, since it is probable that in this way we might learn the significance of the extraordinary variety in the colouring of ornamental flowers.

The typical flowers of cottage gardens—lupins, snapdragons, stocks—are multicolored and seem at first sight to follow no hereditary rules at all. But Mendel saw that by investigating beans—harder than peas to keep track of, but not quite as capricious as some other plants—he (or others) could find the underlying principles even in the hardest cases. Thus we see him edging step-by-step from the simplest cases to the most difficult, exactly as any modern philosopher of science would recommend.*

Here, then, is Mendel's contribution to the science of genetics. He hypothesized that characters are conveyed from generation to generation by individual Mendelian factors (which we now call genes). The parents possess two complete sets of such factors, the gametes contain only one set, and the genes contained in the gametes are combined in the next generation to form an individual who resembles both parents but is uniquely different from

*Mendel saw another reason that flowers in cottage gardens are so multicolored: they are planted close together and interbreed, so that most of the offspring are in fact hybrids. As he says, "It is only the Leguminosae, like *Pisum, Phaseolus,* and *Lens* [the lentil], whose organs of fertilisation are protected by the keel, which constitute a noteworthy exception."

either. It makes no difference whether any one gene is inherited from the male parent or the female: they both behave the same. If an individual inherits a different kind of gene from each of its parents for any one character, then one gene may be dominant, and the other recessive. As a rule (though not invariably), each gene is inherited independently from all other genes, so that, for example, a pea may have seeds that are round and yellow, wrinkled and yellow, round and green, or wrinkled and green.

Those are the basic Mendelian ideas. Clearly, though, Mendel already appreciated some of the necessary refinements. He saw that, sometimes, two different characters may seem to travel together, as white seeds and white flowers go together in garden peas. Clearly, too, as he shows with his "minor experiments" with *Phaseolus* beans, he appreciates that some characters might be conferred by consortia of genes working together. He was correct in this and recommended that the point should be followed up, though he never had sufficient time and resources to do so himself.

As we will see in the second half of this chapter, these ideas have been refined and expanded throughout the twentieth century, and with the refinements and expansions they form what is known as classical genetics. Grasp Mendel, in short, and the rest falls naturally into place, intricate though it may sometimes be. All genetics is footnotes to Mendel.

Mendel was indeed a genius, but he was not the only outstanding intellect of the mid–nineteenth century, nor the only one interested in heredity. How did he manage to get it all right— effectively define the subject in a few brief growing seasons, with a few pots of homely garden plants—when other great thinkers were at sea? It is probably fatuous to pit Mendel against Darwin, but if we were to do such an invidious thing, I would agree with the majority—that Darwin was the greater—for nothing quite compares with the majestic sweep of his vision. Yet it was Darwin who lamented in 1859 that ". . . no-one can say why the same peculiarity in different individuals . . . is sometimes inherited and sometimes not so: why the child often reverts in certain characters to its grandfather, or other much more remote ancestor . . ."; and it was Mendel who, by 1865, had said all that really needed to be said to resolve such issues and a great deal more besides. (There is no excuse for Nathusius, who observed in 1872 that ". . . the laws of heredity have not yet fallen from the tree of knowledge," because by then they very definitely had.) So what did Mendel bring to bear that the other great minds did not?

Well of course there was his background: he was born into a gardening community, was brought up among people—both practical gardeners and intellectuals—obsessed with the problems, and spent his professional life in a monastery that was, in effect, a research station, and was a member of a society (the Brno Natural Science Society) that seemed to include most of the great biologists of mainland Europe. The same kinds of point apply to Darwin, however. True, Darwin was not born among peasant farmers, but he did consort with gardeners and pigeon fanciers, trying to pick up the same folklore that was in Mendel's bones.

The key difference surely lies in Mendel's approach. First and foremost he was by inclination a physicist and mathematician. As a physicist it was his instinct to seek the simplest possible rules that lie behind all natural phenomena, and during his formal education at Vienna and elsewhere he was surrounded by people, like Doppler, who emphasized the need for this. As a physicist, too, he was inclined to think of digital mechanisms: in this instance, the notion that heritable characters might be conferred by corresponding, discrete "factors," and not through the messy mixing of inks and semens. As a mathematician, he knew that if he wanted to make sense of anything in nature, he had to quantify and to apply statistics. Again, many of his teachers specifically stressed this. Darwin, by contrast, was innumerate, as he confessed in a letter to a friend when he left his school at Shrewsbury in 1828: ". . . my noddle is not capacious enough to retain or comprehend Mathematics—Beetle hunting and such things, I grieve to say, is my proper sphere" (quoted from *The Salopian Review,* 1999, p. 33).

Darwin was more than a beetle hunter, of course. He was the greatest naturalist of all time. But herein lies the final and most important distinction. Naturalists seek out and admire the variety of nature. No one has ever known more about nature's variousness than Darwin. But this is the trouble. For every simple case he knew, of every phenomenon, he also knew a hundred exceptions, and as a naturalist, a glorifier of diversity, he could not resist the exceptions. Mendel knew that to get to the core of heredity he had to put aside the exceptions and identify the simplest possible cases, explore them in full, and then work outward. Darwin's broad vision was perfect, indeed indispensable, for the grand task he set himself; that of teasing out the underlying mechanism of evolution. Mendel's more "reductionist" and numerical approach was ideally suited to the job that he undertook. They complemented each other perfectly, and we have inherited the fruits of their complementary genius. Mendel knew about Darwin, and understood him. Darwin, evidently, did not know about Mendel (though he did recommend a book to a friend which contained references to Mendel's work).

But there is one final ghost to lay to rest.

R. A. FISHER—AND WHY MENDEL DID NOT CHEAT

Sir Ronald Fisher was one of the great intellects of the early twentieth century: a mathematician turned biologist (not dissimilar from Mendel, though a better mathematician). As a third-year undergraduate in Cambridge in 1911, he suspected that Galton's ideas of "biometrics" could and should be reconciled with Mendel's genetics, though most biologists seemed to imagine they were naturally opposed; by 1918 he had shown how the two sets of notions could indeed be aligned. He made it all clear in 1930, in *The Genetical Theory of Natural Selection.* Fisher was, in short, a key figure in formulating the "modern synthesis"—the harmonization of Mendel's ideas with Darwin's—as discussed in chapter 6; he remains one of the principal scholars of Mendel's work and one of the elite who truly understood, and could improve upon, Mendel's statistical deliberations.

It was a shock, then, when Fisher published an essay in 1936 entitled "Has Mendel's Work Been Rediscovered?" in which he argued that Mendel's wonderfully clear statistics—so clear that anyone can see the 3:1 and 9:3:3:1 ratios that underpin his theory—were, in fact, too good to be true.

The point can be demonstrated by tossing a coin. On average, if there is nothing wrong with the coin or with the toss, heads and tails should occur with equal frequency: 50 percent each. We all know, however, that if we tossed the coin only twice, then we are quite likely to get two heads or two tails. If we tossed it half a dozen times, we might get half a dozen heads, or half a dozen tails. English captains of cricket tend to call the same every time—it's as good a strategy as any—and in recent years have proved wonderfully adept at losing half a dozen tosses in a row. If we tossed the coin 100 times, however, we would be surprised to get 100 heads, or 100 tails. We would probably be a bit surprised if the ratio of one to the other was greater than, say, 70 to 30. (I believe, in fact—in contrast to what most scientists are inclined to say—that human beings are very good at assessing odds intuitively. Most of us would not be surprised if four tosses produced three heads and one tail; but we would be surprised if 100 tosses produced 75 heads and only 25 tails.)

In short, although there is a 50:50 chance of producing heads (or tails) on any one throw, we would not expect to achieve an exact 50:50 ratio in any one series of throws. But we would expect the ratio to move closer and closer to 50:50 the more throws we made. If we made a million throws, we would not expect the discrepancy to be more than a few thousand—a very small proportion of the whole million.

Most of us, as I say, know this intuitively; the more throws you make, the closer you are likely to get to the theoretically expected 50:50. Professional statisticians, however, are able to be far more precise. They can say that the chances of getting 6 heads in a row are 1 in 64 (1 in $2 \times 2 \times 2 \times 2 \times 2 \times 2$), whereas the chances of getting 100 heads are—well, very small indeed (1 in $2 \times 2 \times 2 \ldots$ etc., 100 times). With a little more manipulation, they can also work out what the chances are of getting, say, a 90:10 ratio, or an 80:20 ratio, or a 99:1 ratio in 100 throws—or 1,000, or a million. Statisticians, in short, can work out what the chances are of getting any particular result from a particular experimental setup.

Mendel demonstrated very clearly that in the second generation of self-fertilized offspring, the plants with the dominant feature outnumbered those with the recessive feature by the ratio 3:1. To be precise, as we have seen, he reported an "average ratio of 2.98:1, or 3:1." But, said Fisher, there is very little chance that Mendel could have achieved ratios that were so close to 3:1 with the relatively small samples he was able to grow in his monastery garden. Indeed, the chances that he could have achieved such fine results are so low as to be incredible. In Fisher's own words: "A serious and almost inexplicable discrepancy has . . . appeared in that in one series of results the numbers observed agree excellently with the . . . ratio which Mendel himself expected, but differ significantly from what should have been expected had his theory been corrected to allow for the small size of his test progenies." Later he wrote: ". . . most, if not all, of the experiments have been falsified so as to agree closely with Mendel's expectations."

So what did Mendel do? Many have speculated since. Perhaps some assistant, in a spirit of helpfulness, massaged the results. Perhaps that well-intentioned but misguided assistant transferred seeds that were somewhat marginal in character from one pile to another, to make the results come out better. Perhaps Mendel was simply too old-fashioned and did not realize that his experiments were actually meant to provide raw data. Perhaps he felt they were for demonstration purposes only and that it was fair game to bend the results so as to make the point more clearly, given that the point was obvious enough. Perhaps, some modern iconoclasts have suggested, he simply cheated. Some, after all, seem very eager to demonstrate that all scientists cheat as a matter of course; that Mendel was a cleric reinforces his status as fair game.

Yet it is hard to doubt for one second that Mendel was scrupulously honest. He was a scientist through and through and knew perfectly well that if data are fiddled, then the whole fabric of science collapses. Everything about his life, and the comments of those who remember him, attests to his honesty. He knew he was presenting his results to the greatest scientists in Europe, and in his paper he stresses how necessary it is to repeat his work. Even a cynic must acknowledge that it would have been risky indeed to cheat in such circumstances. Neither did Fisher ever doubt Mendel's probity. For example, in a paper of 1955, Fisher praises Mendel's excellence (as he had done throughout his life) and never mentions the caveats he raised in 1936. Perhaps he was simply tired of the subject. Perhaps he was irritated that others had used his arguments to cast a shadow on Mendel's work and genius.

I do not claim to know the answer. If Fisher couldn't solve it, I certainly am not going to try. I commend, however, an excellent paper by Teddy Seidenfeld of the Department of Philosophy and Statistics, Carnegie Mellon University. Seidenfeld makes two main points. First he suggests that, in truth, Fisher's analysis is unjust. The essence is that Mendel anticipated the kinds of objections that Fisher, much later, homed in on and designed his experiments in ways that overcame them. This subtlety of design is not directly evident from Mendel's 1866 paper, but it is there if you look. This, says Seidenfeld, answers some of Fisher's criticism, though not all of it.

So what of the rest? Intriguingly, the answer here may be botanical. As we will see later, each "germ cell" gives rise to four gametes, which, in the case of flowering plants, means four eggs and four grains of pollen. In a hybrid plant, each pollen grain has an equal chance of containing a dominant gene (for any particular character) or a recessive gene (for the same character). Because the pollen grains are produced in fours, there would be a checkerboard pattern of grains on the surface of the anther: dominant-gene grains alternating with recessive-gene grains. Peas are self-fertilizing, which is why Mendel chose them. Mendel's reported results also show that his particular peas were low-yielding: he obtained only 30 seeds per plant, while most growers would expect 100 or more. Seidenfeld suggests that in low-yielding plants of the varieties that Mendel grew, pollination would *not* be quite random, because the pollen grains were arranged in the checkerboard fashion on the anther. If fertilization was not random, then the results would be biased. In fact, the results would be more or less as Mendel reported them.

Seidenfeld points out that his hypothesis could be tested by repeating Mendel's experiments precisely: growing the same varieties under conditions in which the yields were poor. It would be good to do this. But even if the peas were to fail the Seidenfeld test, I would still find it impossible to believe that Mendel was in any way to blame. His results may not seem plausible as they stand. But the idea that he cheated is even less plausible.

However, I have said enough about Mendel to make my point—that he was great, that everything since is extrapolation. It is not possible to discuss all the ideas that have emerged since in the same detail. Instead, I shall rush with indecent but I hope useful haste through the main ideas of classical genetics as they unfolded in the twentieth century. Some of them extrapolate from Mendel; some merely provide vocabulary for his ideas.

A LIGHTNING OVERVIEW OF TWENTIETH-CENTURY CLASSICAL GENETICS

Mendel's work was effectively lost to view from the time he presented it in his lecture of 1865 until 1900, when Carl Correns in Germany, Hugo De Vries in Holland, and Erich von Tschermak in Austria independently brought it to light. The science of genetics, then, properly begins with the twentieth century. Mendel *might* have lived to see his work revived and vindicated—he would have been 78 in 1900—but, sadly, he did not.

Mendel was eager to show that the rules he had found in peas apply elsewhere, and others soon found the necessary proof. In 1902, William Bateson in England showed from his work on fowl that Mendelian rules apply to animals. Bateson also coined the term *genetics* in print in 1905, although as we have seen (in chapter 3), the German *"genetische"* had appeared in 1819—albeit, of course, with a pre-Mendelian meaning. In 1909, Bateson's friend Wilhelm Johannsen proposed the term *gene* to replace the somewhat clumsy "Mendelian factor." After that, things moved apace. Archibald Garrod in England began the process that was to link the study of genes, "genetics," to biochemistry, and hence to molecular biology—in other words, the study of what genes *are* and how they actually work. In the rest of this chapter, however, I will look at the study of how genes behave to produce different patterns of heredity: the study known as classical genetics. For brevity, the account takes the form of an extended glossary of notions.

Beads on a String

When scientists don't really know what things *are*—or even when they do—they like to build "models" of them: sometimes simply in their heads, and sometimes literally. In classical genetics, genes are viewed in a quasi-literal light.

At the end of the nineteenth century, microscopists became aware that cells contain "chromosomes": peculiar threadlike structures that appear when cells are preparing to divide (but disappear again when division is complete and the cell nucleus re-forms). The idea grew that these chromosomes are concerned with heredity. Following the rediscovery of Mendel's work at the start of the twentieth century, biologists began to think of hereditary factors as discrete entities, that is, as "genes." What more natural, then, than to think of chromosomes as strings of genes, like beads on a necklace? Now, of course, we know that each chromosome consists of a giant "macromolecule" of DNA. But in truth the genes *are* positioned end to end along the DNA; so the rough-and-ready model that sees chromosomes as necklaces and genes as beads works very well for many purposes. We can say that each creature has as many strings of beads as it has chromosomes.

Indeed, we can hang a great many useful concepts with their accompanying vocabulary on this model. For example, you and I are both human beings, but we are not identical twins. Because we are both human, in a sense we have the same genes. But if our genes were literally identical in every detail, then we would be identical twins. So what's going on?

Well, we can simply say that what you and I have in common as human beings is *positions* where particular genes fit on our chromosomes. Each position is known as a locus (plural, loci). But each gene, in any one locus, *might* come in two or more different forms. Each form—each variant of any one gene—is known as an allele (an abbreviation of Bateson's proposed term, "allelomorph"). So you and I have the same loci, but we have different alleles in at least some of those loci. In fact, we have about 34,000 loci each, and the genes that are positioned on many of those loci do indeed occur in two or more alleles, some of them in many different forms. Thus the *combination* of alleles that each of us possesses is unique (except in the case of identical twins), and the number of possible combinations, when you have different alleles at any or many of the 100,000 loci, is effectively infinite.

Human beings are genetically different from, say, oak trees because we have different loci from them; but human beings differ from other human beings, and oak trees from other oak trees, because they have different versions of particular genes—different alleles—in at least some of the loci. In fact, since human beings and oak trees shared an ancestor in the distant past (around a billion years ago), we should not be entirely surprised to find that *some* of the genes that we possess are the same as some of those in oak trees. But they are quite likely to be positioned differently, on different loci.

The total apportionment of genes in any one individual is called its genome, and the study of all the different genes in any one species (or individual) is called genomics. Genes that exist in two or more different allelic forms are said to be "polymorphic." Populations of creatures that exist in two or more different forms are also said to be polymorphic. *Polymorphic,* after all, simply means "many forms," so there is no reason not to apply the term in several different contexts. A gene that exists in only one version in any one population is said to be "fixed." This is a breeder's term. It means that no matter what you do, the feature ("character") brought about by that gene cannot vary, since there is only one version of the character in the population. The total apportionment of alleles contained in all the genomes in all the individuals in any one breeding population is referred to as the gene pool, although it would be more accurate to call it the allele pool. The total quantity of different alleles within the gene pool is called the genetic variation, though again, it would be more accurate to speak of allelic variation. Scientific vocabulary isn't always as sharp as it ought to be.

Finally, Mendel was keen to emphasize that any character could be recombined, in the offspring, with any other character: a principle sometimes called independent assortment. But when two genes are located close to each other on the same chromosome, they are often passed on as a duo. Then the two are said to be "linked," and the overall phenomenon is called linkage. Indeed, classical geneticists have been able to work out whether two genes are or are not on the same chromosome—an exercise in "gene mapping"—by seeing whether or not they tend to be inherited together. Then they could figure out just how close any two linked genes were to each other by seeing how often they were passed on as a duo.

Darwinian natural selection affects the position of genes on the chromosomes, just as it affects which genes actually survive. Thus we commonly find that genes tend to be linked when they all contribute to the same physiological system. The most spectacular example of linkage is provided by the "hox" genes, which control the total layout of an animal body. The different hox genes are closely linked.

Homology, Homozygosity, Heterozygosity, and Much That Follows

Mendel inferred that each individual adult plant contains two sets of heritable factors, or, as we would say, two sets of genes. In fact, as can clearly be seen under the microscope, each cell of each plant or animal contains two complete sets of chromosomes, each one of which may be envisaged as a string of genes. One set is inherited from the mother, the other from the father. The two sets are said to be homologous. Any one gene at any one locus on any one chromosome is also said to be homologous with the equivalent gene on the same locus on the homologous chromosome.

With these very simple ideas, derived from Mendel, we can tighten the vocabulary that Mendel applied to his own experiments and clear up much of the confusion that existed among breeders and biologists before Mendel.

Sometimes two homologous genes on two homologous chromosomes are identical. This is because the individual has inherited the same version of that gene—the same allele—from each parent. Then the individual is said to be "homozygous" for that particular gene. An individual may, however, have inherited two different versions of the same gene from its parents. Then it will have one allele on one chromosome, and a different allele on the equivalent locus on the homologous chromosome. Then the individual is said to be "heterozygous" for that particular gene. For Mendel and his contemporaries, the term *heterozygous* would be essentially synonymous with the term *hybrid*.

Now we can apply these notions and terms to Mendel's experiments. Round-seededness in his peas was conferred by one kind of allele; wrinkled-seededness was conferred by a different but homologous allele. Individual plants could be homozygous for either of the alleles, containing either two round-seed alleles or two wrinkled-seed alleles. All the gametes that come from a round-seeded homozygote are bound to contain a round-seed allele, so all the offspring produced by two round-seeded homozygotes are bound to have round seeds. Similarly, all the offspring of two wrinkled-seeded homozygotes are bound to have wrinkled seeds.

However, we must now distinguish between "genotype" and "phenotype." *Genotype* refers to the genes that an individual contains; *phenotype* refers to the individual's appearance (or chemistry, or behavior). A pea that is heterozygous for the gene that determines seed shape—an individual with a round-seed allele on one chromosome, and a wrinkled-seed allele on the homologous chromosome—clearly has a different genotype from one that has a round-seed allele on both homologous chromosomes. But because the round-seed allele dominates the wrinkled-seed allele, the heterozygote ends up with round seeds. Phenotypically, the heterozygote is identical with the round-seeded homozygote. In other words, you cannot tell just by looking at a round-seeded pea whether it is homozygous or heterozygous. You can tell only by breeding it and looking at its offspring. Homozygous round-seeds are "true breeding" because all their offspring are bound to inherit two copies of the round-seed allele. But heterozygous round-seeds are not true breeding, because they produce some gametes that contain the wrinkled-seed allele, and if two of those wrinkled-seed alleles get together, they will produce a wrinkled-seeded offspring. Of course, a wrinkled-seeded plant is bound to be homozygous for the wrinkled-seed gene. If it was heterozygous—containing a round-seed allele as well as the wrinkled-seed allele—it would have round seeds, since the round-seed allele dominates the wrinkled-seed allele.

The notions of homozygosity and heterozygosity, and of dominance and recessiveness, take us into vast and intriguing territory. For sometimes genes change: that is, they "mutate" (sometimes under the influence of identifiable "mutagens," and sometimes, it seems, simply through imperfect copying). If they did not mutate, there could be no polymorphism. Genes may exist in several or many different allelic forms because, at some time in the past, there has been mutation.

However, mutations are chance events, and only a minority of such events can produce advantages. Most mutations are harmful, or deleterious. Any one population is bound to contain at least some mutant, "deleterious alleles," which are sometimes perfunctorily referred to as bad genes.

Deleterious alleles sometimes simply kill the creatures that contain them. At least some deaths that occur in the womb or the egg must be caused by such lethal mutations. Sometimes, however, deleterious alleles persist within the gene pool. But how are they able to persist, if they are harmful or even lethal? For three different kinds of reason.

First, and most commonly, many persistent deleterious alleles are recessive. Individuals that are homozygous for such alleles suffer their ill effects. But in individuals that are heterozygous for such a gene, the deleterious allele is silent; the homologous normal allele dominates. In modern parlance, we say that the dominant allele is "expressed" and the recessive one is not. Natural selection would generally eliminate genes from the population if they killed their possessor every time they appeared. But natural selection cannot eliminate a potentially harmful gene if the harmful gene is recessive and so—usually—is simply unexpressed.

Thus potentially harmful alleles lurk in the gene pool, within the genomes of heterozygotes, who are said to be "carriers" of the particular gene. However, they *do* cause disease when they are inherited from both parents, producing an offspring who is homozygous.

Throughout the course of the twentieth century, medical scientists have identified at least 5,000 different deleterious alleles in human beings that, when present in a double dose in homozygotes, cause obvious and sometimes fatal diseases, commonly known as single-gene disorders. The "frequency" (another technical term) with which particular deleterious alleles occurs varies from population to population. Among north Europeans, the most common single-gene disorder is cystic fibrosis. Among people of African descent, by far the most common is sickle-cell anemia. Some Mediterranean peoples and Southeast Asians commonly suffer from a family of similar inherited anemias known as thalassemias. Ashkenazi Jews are particularly plagued by Tay-Sachs disease, and so on.

With modern medicine, sufferers from these and many other diseases can be helped to live fulfilled lives, and may indeed live to have children of their own. Without such therapy, most would die before reaching sexual maturity. Some people who might loosely be called eugenicists have suggested that it is a bad thing for the future of the human species to enable, or allow, people suffering from single-gene disorders to breed. This, they argue, simply allows the "bad genes" to spread, and in the end would be harmful for the human species.

Fortunately, this argument is fatuous as well as cruel. First, we may simply observe that ever since these bad genes first appeared in the human gene pool—which in some cases may have occurred hundreds of thousands of years ago—natural selection *has* been eliminating the homozygotes. Without modern medicine, the sufferers often could not survive. Yet the bad genes persist. Clearly, mere elimination of homozygotes does not work. Why not? Because for any one allele, the heterozygotes are bound to outnumber the homozygotes many times over, so the mere elimination of homozygotes makes virtually no difference to the total frequency.

We can see the point by reference to cystic fibrosis. The deleterious allele is present—remarkably—in about 1 in 20 Caucasian people; that is, 1 in 20 is a heterozygous carrier. There is thus a 1 in 400 (20 × 20) chance that two carriers will mate and produce children. There is only a one in four chance that such a child will inherit two copies of the cystic fibrosis allele and so suffer from the disease. Thus the incidence of cystic fibrosis in the Caucasian population is 1 in 1,600 (1 in 400, divided by four). If we cruelly prevented people with cystic fibrosis from having children, we would be tackling only a very small proportion of the people who actually carry the gene: 1 in 1,600, as opposed to 1 in 20. If a deleterious allele is present in only 1 in 1,000 people, then the incidence of homozygotes would be 1 in 4 million. That is, the rarer the allele, the greater the discrepancy between the number of heterozygous carriers and the number of homozygous sufferers.

Each human being is estimated to carry an average of five deleterious alleles that could cause disease if any two carriers produced offspring. So to eliminate *all* deleterious alleles from the human gene pool, we would have to wipe out the species. Adages involving babies and bath water spring to mind. In general, however, we need have no fears that humanitarian medicine will lead to the decline of the human race.

The second reason apparently deleterious alleles may persist is that at least some of them—and probably more than we know—do bring some benefit when present in the heterozygous form. Recessiveness and dominance are not necessarily absolutes. Sometimes the recessive gene is expressed to some extent: the degree of expression is known as the penetrance. Sometimes, therefore, the heterozygotes are a little different, phenotypically, from the homozygotes. And sometimes the heterozygotes gain an advantage.

Such reasoning explains why the sickle-cell and thalassemia alleles are as common as they are. The genes affect the form of the hemoglobin, the red pigment in the blood that carries oxygen. People who possess two copies of the "normal" gene have normal hemoglobin and do not suffer from inherited anemia. But people with normal hemoglobin are extremely susceptible to malaria, if they live in a malarial region. People who are heterozygous for the sickle-cell allele or one of the thalassemia alleles, however, are less susceptible. When the malaria parasite invades their red blood cells, the cells collapse and the parasite dies. In Africa, and to a lesser extent in the tropics worldwide, malaria is rife. In such areas, people who are heterozygous for sickle-cell anemia or thalassemia have some advantage: a built-in form of protection against a parasite that the immune system seems virtually powerless to ward off. To be sure, the homozygotes tend to die young from inherited anemia, but this disadvantage, cruel as it is, is outweighed by the protection afforded to the heterozygotes. And of course, the heterozygotes greatly outnumber the homozygotes. So natural selection favors the sickle-cell and thalassemia alleles in malarial regions. The alleles then are more common than might otherwise be expected. Of course, if the sickle-cell allele became common, then natural selection would come down hard on it, since then too many people would die from inherited anemia. Thus the level of the sickle-cell allele, while higher than expected, is still low. This is called frequency-dependent selection: natural selection favors a particular allele until it becomes too common (too frequent), at which point it is selected against. Frequency-dependent selection produces "balanced polymorphism": different alleles of the same gene are maintained in a more or less constant ratio to one another.

Both the above arguments, however, refer to deleterious alleles that are recessive. A few "bad genes" persist in the human population even though they are dominant. The most famous example is the one that causes Huntington's chorea, a dreadful disease that inexorably incapacitates the nervous system, and of course curtails life. How can it possibly persist? Because it is not expressed until late in life—after the individual who possesses it has reproduced and passed the gene on to the next generation. Thus natural selection sometimes allows deleterious mutant genes to persist by encouraging recessiveness, so the gene is not normally expressed, and sometimes, as in Huntington's, simply by postponing its expression.

The example of cystic fibrosis and sickle-cell anemia (and any one of 5,000 other examples) are specific instances of a much broader generalization, that far more often than not, heterozygosity is a good thing. Darwin noted the phenomenon of hybrid vigor: creatures that are outbred often tend to be more vigorous than those that are more inbred. We noted in the last chapter the case of "mulattoes" or "half-castes" who may have been despised historically but often seem especially bright and/or athletic. In this chapter we noted Mendel's outbred peas, which tended to be taller than the taller parent, even when the other parent was a dwarf. From the discussion of human single-gene disorders, we can easily infer the reason. Although only a few of our alleles are obviously deleterious, some do not do their job as efficiently as others. If we inherit two doses of an allele which, though not actually *bad,* is not the best, then we will not be as well off as we would if we inherited two different alleles of the same gene, one of which might be a superior type. Thus *in general,* heterozygosity is favored. "Hybrid vigor" is more generally known these days as heterosis.

This is why some plants have evolved specific "mating barriers." Those prevent inbreeding—mating with other individuals that are genetically too similar—which would lead to excess homozygosity. Some plants, however, are adapted to inbreeding and can tolerate high degrees of homozygosity. We must assume that such plants do not have too many deleterious alleles in their gene pool. Peas are of this type. So is wheat, and so is barley. Corn and millet, by contrast, are natural outbreeders and tend to be severely compromised if they are allowed to become too homozygous. But as we saw from Mendel's extratall crossbred peas, even natural inbreeders benefit from a little extra heterozygosity now and again. However, as discussed in chapter 2, before the mechanisms of homozygosity and heterozygosity were understood—indeed before the concept of the gene was formulated—the caprices of inbreeding and outbreeding were extremely confusing.

Sex Linkage

Darwin asked why inherited characters sometimes appear more in one sex than in the other. Among such characters are hemophilia, which is far more common in boys than in girls (and plagued many of the males in Queen Victoria's family). The reason is that in mammals and birds and many other animals (though not, for example, in all reptiles), the sex is largely determined by the possession, or non-possession, of particular chromosomes. Female mammals have two X chromosomes, both of which are much the same size as the other chromosomes; male mammals have one X chromosome and one Y chromosome, and the Y is diminutive. If a deleterious allele occurs on the X chromosome, then in females it is likely to be balanced by a normal homologous allele on the other X chromosome. But if the same allele occurs on the X chromosome in a male, then it may well find itself unopposed by any homologous allele, since the diminutive Y chromosome may not contain any homologue. The allele that leads to hemophilia in humans occurs on the X chromosome, and has no homologue on the Y. So whenever it occurs in males, it leads to disease. Female hemophiliacs are possible but are produced only when a hemophiliac male mates with a carrier female: a rare event.

Incidentally, the X and Y chromosomes are called sex chromosomes, and the rest (which are the same in both male and female mammals) are called autosomal chromosomes.

Polygenic and Pleiotropic

Mendel deliberately homed in on characters that clearly had a simple hereditary basis, what twentieth-century geneticists call single-gene characters. However, as Mendel well knew, only a minority of characters are determined so simply, by single genes. Most are influenced by whole batteries of genes working in concert. Such characters are said to be "polygenic." In humans, the general character of the hair and skin are each influenced by batteries of genes.

Also, it has become apparent in the twentieth century that many genes—perhaps most—affect more than one character, and that the characters that may be affected by one gene may be very different from each other and have apparently very little connection. Thus the deleterious allele that causes cystic fibrosis affects, principally, both the lungs and the pancreas. Such many-actioned genes are said to "pleiotropic."

If to Mendel's simple rules of heredity we now stir in the notion that many (perhaps most) characters are polygenic in origin, and that many (perhaps most) genes are pleiotropic in action, we can quickly see intuitively how those simple rules, when applied in practice, can lead to endlessly complex results. This does not mean that Mendel was a simple man, unable to grasp complexity. It means instead that he was a genius, able to see the simplicity that lies beneath the astonishing complexity of nature.

The Matter of Ploidy

Finally, we noted earlier—Mendel implied as much—that creatures like us, and oak trees and fungi (though not *all* creatures), contain two sets of chromosomes in each of their body cells. Such creatures are said to be "diploid." Gametes, however—and some adult organisms—contain only one set of chromosomes and are said to be "haploid." In general, the number of sets of chromosomes in an organism is referred to as its "ploidy."

However, before a diploid cell divides, it doubles each of its chromosomes. After this doubling has occurred, a normally diploid cell is, in fact, "tetraploid": that is, it contains four sets of chromosomes. But in some organisms—including a great many plants—the cells are normally tetraploid (and after the chromosomes have divided, as a warm-up to cell division, they are octoploid). Organisms with more than two sets of chromosomes are said to be "polyploid."

The ability of plants to become polyploid enables them to produce new species apparently out of the blue. When creatures of different species mate, they often cannot produce viable, fertile hybrids because their chromosomes are incompatible. The chromosomes need to cooperate with their homologues to produce gametes, and if the two sets of chromosomes inherited from the two parents are different, then no one chromosome has a convincing homologue with which to team up. Sometimes, however, a haploid gamete from one kind of plant combines with a haploid gamete from another kind of plant, related but different, to produce a diploid hybrid offspring; then all the chromosomes in that diploid offspring double *without* subsequent cell division, to produce a tetraploid organism. By doubling, each chromosome acquires an instant homologue. So the tetraploid offspring of two diploid parents can produce gametes and so be fertile. But the tetraploid offspring cannot now interbreed with either parent, and so has formed a new species. Thus rutabagas are perfectly fertile tetraploid hybrids formed by the mating of turnips and cabbages, although turnips and cabbages are different species. Speciation by tetraploidy seems to be common in plants, both in the wild and in cultivation. It can be seen in whole series of wild aloes, for example, which are succulent plants that grow in Africa. The domestic European potato is a tetraploid, produced by doubling the chromosomes in a South American diploid ancestor. Bread wheat is a hexaploid hybrid, containing six sets of chromosomes, the complete genomes of three wild ancestral grasses. Animals, however, are clearly much less tolerant of polyploidy.

It is possible to go on discussing such ramifications more or less indefinitely. Twentieth-century geneticists were busy, and many were exceedingly brilliant. I maintain, however, that *most* of the principal ideas are outlined in the 30 or so preceding paragraphs, and that if you can see the point of them then you would be able at least to get a handle on the vast corpus of twentieth-century genetic scholarship. I also maintain that most of the ideas outlined above extrapolate quite readily from the ideas of Mendel. But at this point the story bifurcates. Some twentieth-century biologists asked what genes actually *are,* and how they work; out of this came the practical crafts of biotechnology, with all its extraordinary present-day power. Others, in a more theoretical and philosophical vein, sought to reconcile Mendel's ideas of heredity with Darwin's theory of evolution, to see where that would lead.

FIVE

What Genes Are

and How They Work

At the start of the twentieth century the newly defined concept of the "gene" was still an abstraction: a "factor," as Mendel had said, that affected the phenotype of a creature—the way it looked and behaved—in particular ways. The whole of "classical" genetics, which came of age in the early decades of the twentieth century, treats genes as abstractions, or at least as beads on a string. Yet early in the twentieth century—thanks to the insight of an outstanding English physician, Sir Archibald Garrod—biologists gained some insight into what genes actually *do:* to wit, they make proteins. The best part of another half century passed before anyone could be sure what genes actually *are:* stretches of DNA. The three-dimensional structure of DNA was made clear in the 1950s, and this insight gave rise to a new discipline of science known as molecular biology. For the past half century, the new molecular biologists have sought to understand how DNA works its magic, with many a surprise along the way.

So now, at the start of the twenty-first century, biologists can still think of genes in abstract terms when it suits them to do so, but they can also think of them as chemical entities, involved in complicated chemistry. It is often appropriate for biologists of many kinds—breeders, physicians, conservation biologists, evolutionary psychologists—to think of genes as beads on strings: classical genetics is alive and well and always will be. Increasingly, however, classical genetics may cross-refer to molecular biology, approaching genes as an exercise in chemistry. The two in combination are powerful indeed. Classical genetics underpins the crafts of agricultural breeding and genetic counseling; molecular biology has generated the technology of genetic engineering. This chapter traces the shift in understanding that took place through the twentieth century and still continues: the perception of genes qua beads to genes qua discrete chemical entities that carry out specific tasks and can be manipulated by genetic "engineers."

FIRST STIRRINGS: THE LATE NINETEENTH CENTURY

The hiatus between Mendel's announcement of his pea experiments in the 1860s and the rediscovery of his work at the start of the twentieth century is regrettable, but there were important developments nonetheless. First, in the 1860s, chromosomes were discovered. Biologists watched living cells dividing. Before a cell divides, the nucleus—the most conspicuous feature of the cell—simply disappears, or, to be more precise, the membrane that surrounds the nucleus breaks down, so that the nuclear contents suddenly become continuous with the surrounding cytoplasm. At the time this happens, a small battalion of strange entities swim into view that will absorb colored dyes, and hence their name: *chromosome* is Greek (or Greek-ish) for "colored body."

Soon it became clear that these chromosomes come in pairs: each one has a homologous partner. It was clear, too, that each species has its own characteristic number and pattern of chromosomes. Human beings, for example, have 46 chromosomes: two homologous sets of 23. Chimpanzees have 48: two homologous sets of 24. Fruit flies have 4 chromosomes. Cabbage has 18: two sets of 9. There is no particular rhyme or reason (or at least, none that is known) that any one species should have a particular number.

Then, in the early 1900s, three American biologists working independently—Clarence McClung, Nettie Stevens, and Edmund Wilson—found that in grasshoppers and other insects the different sexes had different chromosomes. *Most* of the chromosomes in the males and the females were the same, but whereas the females had two X chromosomes—the name X was conferred by Nettie Stevens—the males had only one. Clearly, the number of X chromosomes correlated with sex. But was it possible, asked McClung, Stevens, and Wilson, that the number of X chromosomes not only correlated but actually *determined* sex? Nowadays we know that in many animals, one pair of chromosomes does indeed determine the sex, although many other factors influence secondary sexual characteristics. These are the "sex" chromosomes. The rest—the majority—are the "autosomal" chromosomes. In mammals, females have two X chromosomes while males have one X matched by a diminutive Y.

Even more exciting, though, was the behavior of chromosomes during cell division, as first described in detail in the 1880s. The division is conducted with military precision. When the chromosomes first appear, as the nuclear membrane breaks down, they appear to be *doubled*. Each of the 46 chromosomes of a human being appears to be shaped like a cross, or like a pair of scissors; each chromosome appears as two distinct entities joined at one point, which is called the *centromere*. Each entity in each doubled chromosome is at this point called a *chromatid*, as if it were a diminutive, a half chromosome. But it is not. Functionally, each so-called chromatid operates as an entire chromosome, and the two chromatids are exact copies of each other. Cell division begins when the centromeres break down and each pair of chromatids are separated. Once this has occurred, the chromatids are called chromosomes again. After the chromatids have been split neatly down the middle, there are two complete sets of chromosomes (two complete sets of 46, in the case of human beings). These two sets now separate (they are pulled and pushed apart by strands of protein in the cytoplasm), and each collection of 46 chromosomes acquires a new nuclear membrane. A cell with two complete nuclei now exists, and the cytoplasm divides between the two nuclei to produce two complete cells.

The method by which the chromosomes divide ensures that both daughter cells finish up with identical sets of chromosomes. This kind of chromosome division is called *mitosis.*

Gametes are produced by a somewhat more elaborate method of cell division known as *meiosis,* but a description of this had to wait until the twentieth century. Meiosis roughly resembles mitosis but is a two-stage process. In the first stage (Meiosis-I) there is no separation of chromosomes: the already doubled structures seek out their homologous partners and cleave together. Then they essentially intermingle: bits of each chromosome interchange with equivalent bits from the homologous chromosome. This interchange of chromosomal material is called crossing over, and it results in a complete, new set of "recombined" chromosomes, each of which contains material from *both* the parent homologues. These two paired sets of recombined chromosomes now separate in the phase known as Meisosis-II, or reduction division. Without further doubling of chromosome material, each pair of chromosomes from each set separates from its homologue. Meiosis results in four cells, each of which is haploid, containing just one set of chromosomes, and each of those chromosomes is a unique new entity, combining genetic material from both parent homologues. Each of the four haploid cells now matures to become a gamete, a mature sperm or egg.

Why would the cell go to such lengths to divide its chromosomes so precisely, unless the chromosomes were supremely important? If they are important, then for what? The great August Weismann suggested in the 1880s that their role might be to pass

on hereditary information. This suggestion is the perfect comple-
ment to Mendel's own observations. Mendel suggested that hereditary information is passed on in the form of discrete particles,
with numerical precision, and in mitosis we can see physical entities dividing as precisely as could be required. Many have said
what a shame it was that Darwin did not know of Mendel's work.
What a pity it was, too, that Weismann was apparently unaware
of it.

When Mendel's work did resurface in the 1900s, biologists did
not immediately perceive how neatly his ideas on hereditary factors jibed with the real, observable physical behavior of the chromosomes. William Bateson, though a pioneer of the new genetics,
seemed positively distraught by the thought that the chromosomes might actually carry the genes. Some scientists like to suggest that science is unswervingly logical, but that is simply not
the case. It requires counterintuitive, wild leaps of imagination,
which in turn require creative genius. It is even less true that the
history of science is unswervingly logical, with brick placed neatly
on brick to create an edifice of undisputable truth. Science is a
human activity, and its history is as messy and perverse as that of
all human dealings.

These were the first stirrings, then, of the understanding of
the physical basis of heredity. There is one more thread in this
story, again drawn from the nineteenth century. In 1869, Johann
Friedrich Miescher, a Swiss biochemist working at Tübingen in
Germany, discovered DNA in pus cells. He didn't know what it
was, of course, and simply called it nuclein (the name *nucleic acid*
was coined much later, in 1889, by another scientist, Richard Alt-
mann). Then, in the 1870s, Miescher studied nuclein in the sperm
of Rhine salmon and concluded that it might be "the specific
cause of fertilisation." Even at this late date it was difficult to differentiate between factors that have to do with fertility and those
that have to do with heredity. Miescher was clearly onto something, however. He perceived, at least, that his nuclein could be

highly significant. But he died in 1895 aged only 51, never know-
ing what his discovery would lead to.

Thus at the start of the twentieth century there were some
solid pieces of information: Mendel's genetic rules, Weismann's insight into the possible significance of chromosomes, and Miescher's
discovery of DNA. Of course it was not obvious at the time how
these threads of information fitted together. Another half century
passed before these key observations of nineteenth-century genetics (or what we now call genetics) were brought properly together.
Yet early in the twentieth century—remarkably early—biologists
gained insight into what genes really do. It would be some decades
before they knew what genes actually *are,* or how they operate at
the molecular level.

WHAT GENES DO

Again, we can begin this part of the story with two key discoveries
from the nineteenth century. First, early in that century, chemists
became fully aware that the chemistry of living things is not completely different, qualitatively, from the everyday chemistry of the
laboratory. There is continuity between the two; between living
and nonliving. The compound urea—a principal excretory product of mammals—was analyzed and found to contain nitrogen and
carbon. Thus organic chemistry was born: the chemistry of organisms, but more to the point—and less restrictively—the chemistry of carbon. This was one of the first serious nails in the coffin
of "vitalism": the ancient notion that held that life requires some
extra vital "spark," quite beyond the bounds of normal chemistry.
(Clearly, life *might* require extra forces, over and above those of
mere chemistry, but even if that were so, it was clear that much of
what living things do can be analyzed in terms of ordinary chemistry. But there was certainly no a priori reason to assume that life
had to partake of some extra vital spark).

Later in the nineteenth century, biologists became aware of catalysts within the cells, known as ferments—later called enzymes—that drive and control the metabolism. In 1897 yet another great German scientist, Eduard Buchner, was studying such "ferments" in yeast. These, after all, were the agents that turned grapes into wine, and malted barley into beer. Buchner ground up the yeast to release the cell contents for further study. He decided to store the yeast juice overnight, and to do this he added sugar as a preservative, just as a cook adds sugar to fruit to make jam. When he came back in the morning, however, the juice of the yeast had fermented the sugar, just as the whole yeast organism was known to do. This was the first demonstration that ferments could function perfectly well even when they were separated from the cells that contained them. Buchner's discovery may be taken to mark the birth of biochemistry—and dealt another blow to the ancient conceit of vitalism.

The idea of the enzyme as the catalyst of metabolism was widely established in the early twentieth century when Archibald Garrod, at St. Bartholomew's Hospital in London, was studying the peculiar disorder known as alkaptonuria. Patients with alkaptonuria have peculiar urine, which turns black on exposure to air. This does not matter in itself, though it is important in diagnosis. What does matter is the arthritis that patients tend to suffer later in life.

Garrod and the energetic William Bateson showed that alkaptonuria runs in families, and is most common in families in which cousins marry each other. Effectively—although the numbers were too small to be statistically significant—they showed that the disease followed a Mendelian pattern of inheritance. Garrod then showed that sufferers from alkaptonuria excrete homogentisic acid (HA), which is the material that turns black in air. HA is present in food, but in people with normal enzymes it is broken down chemically before excretion. In a lecture to the Royal College of Physicians in 1908, Garrod suggested that alkaptonuria was an "inborn error of metabolism." Later, he described three more such "inborn errors."

Garrod hypothesized that the failure to break down HA must be due to a defect in a particular enzyme. Since the disease seemed to have a simple genetic basis—that is, the pattern of inheritance suggested that a single recessive gene was at work—he deduced that the failure in the enzyme must be due to a defect in the gene. Thus he coined the brilliant adage "One gene, one enzyme." This truly was a wonderful insight, a resounding inference from a simple observation that was worthy of Mendel himself.

Garrod's adage was confirmed in the 1940s by George Beadle and Edward Tatum at Stanford University in California. They worked with the orange bread mold, *Neurospora crassa,* which normally can live on an irreducibly modest diet of inorganic salts, sugar, and the vitamin biotin. Beadle and Tatum irradiated the *Neurospora* with X rays and then crossed the resulting mutants with healthy *Neurospora* that had not been irradiated. The offspring could no longer live on the simple diet. They needed extra nutrients to survive. Beadle and Tatum inferred that the genetic mutations caused by the X rays had robbed the mold of its ability to produce the enzymes needed to synthesize an adequate diet from simple ingredients. Garrod studied a naturally occurring genetic defect in humans, and Beadle and Tatum studied an artificially induced defect in a fungus, but they reached the same conclusion: one gene, one enzyme.

This seminal observation was then extended. Chemically speaking, all enzymes belong to the grand class of organic compounds known as proteins. Many thousands of proteins function as enzymes. Others are primarily structural: all cell membranes consist of proteins interwoven with fats. Fingernails, hair, and muscles are proteins. Other proteins function as hormones, including insulin. The pigments within the blood that carry oxygen—hemoglobin in vertebrates and some other animals, hemocyanin in mollusks and others—consist mainly of protein. Antibodies, which the body produces to ward off anything foreign, including pathogens, are proteins. Proteins, in short, run the show; they form the structure, they direct the metabolism, they run messages, they form the defensive forces. We may see the genes as the central directors of the cell's activities, but the proteins are the players—the operatives.

In the 1950s the great American chemist Linus Pauling made an observation directly analogous to that of Garrod. He perceived that sickle-cell anemia is caused by a defect in hemoglobin, and that the disease is inherited, apparently in a simple Mendelian fashion (in so far as can be inferred when the number of offspring is too low for rigorous statistics), so that it is probably brought about by a single Mendelian gene. So just as Garrod had inferred that a defective gene produces a defective enzyme, Pauling surmised that a defective gene leads to defective proteins in general. Garrod's "one gene, one enzyme" became "one gene, one protein."

"One gene, one protein" should not be taken absolutely literally. Many proteins—at least in their finished form—require several genes working together. Sometimes different genes make different parts of a protein; often oligosaccharides (middle-size sugars) or metals need to be added before a protein is fully functional. And although proteins largely fold themselves into their finished shape, they are often given a helping hand; the helping hands are typically supplied by other proteins, produced by other genes. Most phenotypic characters, too, are polygenic, requiring many different genes; and most genes are pleiotropic, affecting more than one character. But the central observation—one gene, one protein—is the simple case that encapsulates the essence. All other cases (albeit the majority) are elaborations.

After Garrod, then, biologists knew what genes do; Beadle and Tatum and then Pauling sewed the matter up beyond reasonable doubt. Until the 1940s, however—when the age of classical genetics was well advanced—it was not at all clear what genes actually *are*.

WHAT ARE GENES?

Biologists tend to say as a form of shorthand that genes "make" proteins, but no one ever supposed this was literally the case. What they do, as was inferred from a very early stage, is provide the code on which particular proteins are based.

But just as proteins are astonishingly versatile in function, so they are correspondingly variable in structure, because their structure determines their function. Structurally, in fact, they are infinitely variable. If genes themselves are chemical entities—and no serious biologist in the twentieth century supposed otherwise—then the code they provide must be as versatile as the things they code for. But what, in all of nature, is as versatile and various as a protein? For a time it seemed that the answer was "Nothing—apart from another protein." Small wonder, then, that biologists in the early twentieth century tended to suppose that genes, too, were proteins. Erwin Schrödinger was a particle physicist and not a biologist; nonetheless he wrote a book in 1944 entitled *What Is Life?* that was highly influential. In it he assumed that genes were themselves made of proteins.

The known biology seemed to support this idea. By the 1940s it was beyond reasonable doubt that the chromosomes are involved in heredity, that they carry the hypothetical abstractions known as genes. So what in practice do chromosomes contain? Chemical analysis revealed only two components: proteins and the stuff that Miescher had discovered in the 1860s, nuclein, alias nucleic acid. Protein was already known to be astonishingly various, but nuclein seemed to be rather boringly uniform. It was reasonable to conclude that the genes resided in the protein component of the chromosomes and that the nucleic acid was simply a helpmate, or perhaps merely provided a suitable environment.

Let us look at proteins more closely and seek the source of their variousness. The properties of a protein—whether it behaves as any of a vast number of enzymes, or whether it can function as a piece of fingernail or a muscle fiber—depend, in the end, on its three-dimensional, sculptural form, which chemists call its tertiary structure. But whatever strange and intricate form a protein finally assumes, it begins life in the form of a one-dimensional chain. The finished protein is like a ship's cable, lying in a heap on the dockside, except that the finished shape is far from random; it is molded as precisely as a Henry Moore sculpture. The chain is folded into the finished form partly by "self-assembly"—it folds the way it does partly because that's the way it is—and partly with the help of subsidiary proteins (known as chaperonins) and other enzymes. In the end, though, however the folding is carried out, the tertiary structure of the finished, functional protein depends absolutely on the "primary" structure of the initial chain.

Of what does the basic chain consist? Emil Fischer of Germany showed in the early 1900s that proteins are strings of subunits known as amino acids. Each amino acid is a molecule in its own right; so a protein "molecule" can properly be called a macromolecule. Since proteins are infinite in structure, you might suppose that there is an infinity of amino acids. Organic chemists, in the laboratory, can indeed make as many different kinds of amino acids as they choose, but nature, it turns out, makes do with about 20. All the many millions of different proteins in all the many creatures on Earth are compounded from that basic 20 or so (most of which occur in the proteins found in human bodies). In fact, the amino acid is to the protein what the letter is to the word. All 415,000 or so English words in the *Oxford English Dictionary* are compounded from just 26 letters, and those same letters, with a few accents here and there, also code adequately for the many thousands of other languages and dialects throughout the world.

The task, then, for the gene, is simply to ensure that amino acids—of around 20 different kinds—are lined up in the right order. The infinity of different forms of protein that are seen in nature (or can simply be imagined in our heads) will then follow naturally.

Let us look again, then, at nuclein, to see if it might be up to the task after all. Miescher himself showed that nuclein was an acid, and further analysis quickly showed that it contained a sugar, which turned out in the 1920s to be deoxyribose. Hence nuclein was given the name deoxyribonucleic acid, or DNA for short. Miescher had shown that DNA is confined to cell nuclei, but in the 1920s, Robert Feulgen of Germany went one step further and showed, by judicious staining, that it was exclusive to the chromosomes. Another, similar nucleic acid then turned up, which in fact was found in the cytoplasm at least as much as in the nucleus. This second acid contained a similar but different sugar, ribose. Hence it was called ribonucleic acid, or RNA.

Still though, we must ask, could DNA provide the variety needed to code for the infinity of proteins? The chemical evidence, at first sight, suggests not. It became clear in the first few decades of this century that the two nucleic acids consist of chains of smaller molecules, just as proteins do: not amino acids, in this case, but *nucleotides*. But whereas there are 20 different kinds of amino acids within protein chains, there are only *four* kinds of nucleotides within DNA or RNA. Furthermore, the different nucleotides all seem much of a muchness. Each nucleotide has three components: a sugar (deoxyribose in DNA, ribose in RNA); a phosphate radical; and a base, which is either a purine or a pyrimidine. The sugars and phosphates are the same in every DNA or RNA nucleotide. The only source of variation lies in the four bases. Each nucleotide of DNA contains either one of the two purines (adenine, A, or guanine, G) or one of two pyrimidines (cytosine, C, or thymine, T). RNA is much the same, but the bases differ slightly: uracil, U, another pyrimidine, is substituted for thymine. How can a DNA chain, with only four different kinds of component, provide an adequate code for a protein that has 20 different kinds of component?

It still doesn't seem to make sense—that DNA, apparently so boring chemically, can provide the code for proteins that are infinitely various. Yet the twentieth century provided a succession of experiments that first hinted and finally demonstrated beyond doubt that DNA is indeed the stuff of genes, and that proteins are merely the products. In the 1920s an Englishman, Frederick Griffith, set the ball rolling with his work on the bacterium pneumococcus. The normal, virulent form of pneumococcus causes disease in mice, but it loses its virulence if it is first killed by heat. There are also nonvirulent strains that do not cause disease at all. But Griffith showed that if nonvirulent living pneumococci were injected into mice together with virulent killed pneumococci, the mice died *even though neither form alone was capable of causing disease.* In other words, the two kinds of bacterium, though each deficient in some way, could combine their strengths and between them reacquire the quality of virulence. There was thus an exchange of "information" between the two kinds, an exchange that, as is now evident, is analogous to sex in animals and plants.

In the 1940s, Oswald T. Avery and his colleagues at the Rockefeller Institute Hospital in New York found that there was a specific material that was transferred from the virulent pneumococcus to the nonvirulent type and could "transform" the nonvirulent type into the virulent type. So what was this transforming material? Avery treated it with a protease—an enzyme that breaks down protein—but it did not lose its ability to transform. Therefore the transforming material was not a protein. Then he treated it with an enzyme that was known to destroy DNA, and the ability to effect transformation was lost. Thus, he concluded, the transforming material was DNA. Avery announced these results in 1944.

The second significant line of inquiry that effectively proved the role of DNA in conveying hereditary information involved viruses of the kind known as bacteriophages, usually shortened to phages. *Phage* comes from the Greek for "eat," and phage viruses do in fact make their living as parasites of bacteria (illustrating Jonathan Swift's adage that "a flea/Hath smaller fleas that on him prey;/And these have smaller still to bite 'em;/And so proceed *ad infinitum*"). In particular, biologists studied phages that multiply within the bacterium *Escherichia coli,* which lives in the guts of animals. Inside an *E. coli* host, a single phage can produce several hundred replicates of itself within 20 minutes, but by the 1940s microbiologists already knew that they can do a great deal more besides. They can also exchange genetic information. Thus, if different mutant phages are allowed to infect a single *E. coli,* it isn't long before *normal* types of phages begin to appear: evidently, the "good qualities" from two or more otherwise defective phages can be combined. The parallels with Griffith's and Avery's pneumococcus experiments are obvious.

Viruses are structurally much simpler than bacteria. By the middle of this century, phages were known to consist of a central core of DNA and an outer coat of protein. In 1952, at Cold Spring Harbor, Long Island, Alfred Hershey and Martha Chase showed that only the DNA of phages enters the bacteria they attack. The protein coat remains on the outside. Thus the subsequent reproduction of entire phages within the bacterial host was orchestrated by DNA alone. Presumably, the exchange of information between phages was also enacted through DNA.

In short, more and more evidence showed that hereditary information—genes—was in fact encapsulated in DNA. So now the question was, How can a molecule that apparently has a rather boring basic structure—being made of only four basic components—provide a code for molecules that have a much more complicated structure? This can be answered only by describing the structure of DNA in more detail. Some biologists, though by no means all, realized at the start of the 1950s that this was one of the key issues of biology. The answer did not come easily.

THE STRUCTURE OF DNA

The general model of DNA that is now accepted was first proposed in 1953 by the English physicist turned biologist Francis Crick and the young American biologist James Watson, who worked together at Cambridge, England. Crick and Watson were the great synthesizers. Their strength lay not in performing key experiments, or in compiling crucial data, but in making sense of the observations of others. Biology needs craftspeople who can manipulate delicate cells and instruments, but it also needs thinkers who can make sense of the data thus provided.

The data from which Crick and Watson synthesized their model came first from chemistry—analytical and physical—and then from attempts to describe and measure directly the physical structure of the DNA molecule, both with the electron microscope and, most important of all, by means of X-ray crystallography.

We have already looked at the early chemical findings: that DNA consists of chains of nucleotides. Two outstanding pieces of chemistry after World War II enriched the picture enormously. First, by 1953 the British chemist Alexander Todd had shown that the nucleotides are joined together via the phosphates. In fact, the phosphates linked sugars in adjacent nucleotides, always joining the number 3 carbon in one deoxyribose to the number 5 carbon in the next one. This observation alone gives a rough insight into the overall form of the finished molecule. At least it shows what shapes the complete molecule *cannot* assume.

Observations elsewhere revealed subtleties in the proportions of the four bases in DNA. Biochemists of an earlier age assumed that all four types—A, G, C, and T—would be present in equal amounts: in other words, that DNA was just a boring polymer, like polyethylene, with long chains of identical units. But after World War II it became clear that the four bases were *not* present in equal amounts. So DNA was not like polyethylene. Furthermore—a highly intriguing finding—the proportions of the four nucleotides differed in DNA from species to species.

In the 1940s an Austrian biochemist, Erwin Chargaff, working at Columbia University, added a further and highly pertinent insight. He showed that, whatever the total amounts of the four bases, the amount of A always equaled the amount of T, and the amount of G always equaled the amount of C.

In addition, twentieth-century scientists were able increasingly to explore the structure of complex molecules directly. The electron microscope, in general use after World War II, showed that DNA molecules were threadlike: many thousands of angstroms long, but only 20 angstroms thick (an angstrom is one ten-millionth of a millimeter). Since individual nucleotides were only 3 angstroms across, DNA was obviously compounded from many thousands of them.

But the crucial technique was X-ray crystallography, developed in England from the start of World War I first by William Bragg and then by his son Lawrence, for which they both received Nobel Prizes. The general idea was to deduce the arrangement of atoms in crystals by firing X rays at them. The atoms scatter the X rays, and from the pattern of the diffractions (as revealed on a photographic plate), the position of the atoms can be inferred, the necessary calculations having been done. Early work was carried out on very simple compounds, such as sodium chloride, but even before World War II, W. T. Astbury of Leeds University in England felt confident enough to begin such work on DNA. It is remarkably easy to extract DNA—this has become a standard exercise in school laboratories—and it manifests in a form like fluffy cotton. But when wetted, it becomes tacky and can be drawn into long threads in which the macromolecules line up in parallel: a regular arrangement that is crystal-like. However, to prepare crystals good enough for X rays is no mean technical feat.

The final phase was begun after 1949 at King's College, London, by Maurice Wilkins and Rosalind Franklin, using DNA prepared in Bern, Switzerland, by Rudolf Singer. Crucially, they showed that the DNA chains are arranged in a helix, with the purines and pyrimidines on the inside, and the sugar-phosphate backbone (more of a scaffold than a backbone) on the outside. They also showed that each purine or pyrimidine occupied 3.4

angstroms, and that each turn of the helix was 34 angstroms deep, so there were ten nucleotide molecules per coil. But the diameter of the helix was 20 angstroms, which was too large if the DNA helix contained only one chain. Perhaps, then, it contained two chains; perhaps (as Linus Pauling hypothesized at one point), it contained three.

There was one further, highly intriguing line of thought—not a piece of evidence but an idea, put forward in 1940 by Linus Pauling and Max Delbrück at Caltech. Perhaps, they said, the replication of genes involved a splitting, with each half of each gene acting as a template for the re-creation of a complementary half.

Finally, at Cambridge in 1953, James Watson and Francis Crick put everyone out of their misery by proposing that the DNA macromolecule in fact consisted of two helical chains that run in opposite directions, "head to tail." This was the famous *double helix*—now the favorite symbol of every institution that has anything whatever to do with molecular biology.

The two DNA chains in the double helix are held together by weak chemical bonds (hydrogen bonds) between the bases that run along the center: adenine in one chain joins to thymine in the other; cytosine in one links with guanine in the other. This explains the ratio observed by Chargaff. One chain thus complements the other precisely: if there is a sequence of bases running CATTG in one chain, it will be matched by GTAAC in the other. In principle, though not of course in detail, this is more or less the same kind of mechanism suggested more than a decade earlier by Pauling and Delbrück.

Watson and Crick described their wonderful vision of DNA in the international journal of science *Nature,* in 1953. Their paper consisted of a mere 800 words: tight as a sonnet. It ended with a supremely disarming sentence: "It has not escaped our notice that the specific pairing we have postulated [between A and T, G and C] immediately suggests a possible copying mechanism for the genetic material."

This was the turning point. Molecular biology and all that comes from it—genetic engineering, DNA fingerprinting, and all the rest—begin here. Crick, Watson, and Wilkins shared the Nobel Prize in medicine in 1962 (Rosalind Franklin having died at the tragically early age of 38, in 1958).

Nowadays, too, the structure of chromosomes is perfectly understood. The long, long double helices of DNA are wound around a central core of proteins known as histones. The histones have several functions, and one very obvious one is to provide a robust scaffold that allows the DNA to be hauled bodily about during mitosis and meiosis. Each chromosome contains just one continuous DNA double helix, wound round and round the central histone core. The chromosomes appear during cell division and disappear at other times because when the cell is not dividing the DNA macromolecules are released from the binding core of histones and spread themselves out throughout the nucleus. They can function much more freely in this spread-out form than when they are packed up tight (*condensed* is the technical term) for the purposes of cell division. When the macromolecules are spread out, they are not visible individually through the light microscope.

So now we know the structure of DNA, and we know that it makes proteins. How can we put these two kinds of information together? How, in fact, does DNA work?

HOW DOES DNA WORK?

Here there are several subquestions. The first, perhaps, is: How does DNA replicate itself? The second: How in practice are the protein chains put together, according to the DNA's instructions? And the third is the one we have posed before: How do the four different nucleotides in DNA code for the 20 different amino acids in protein?

Pauling and Delbrück had already guessed the principle of DNA duplication, and Watson and Crick hint at it in the last line of their *Nature* paper. In practice, one strand of DNA splits away from its partner, and from the moment the splitting starts, nucleotides within the surrounding medium begin to line up with their complementary opposite numbers. As they do so, they are joined one to another by an enzyme, *DNA polymerase.* A point crucial to the later development of genetic engineering is the extent to which the assembly and disassembly of DNA depend upon teams of enzymes, of which DNA polymerase is one.

That DNA really does duplicate in this way was first shown experimentally in a highly ingenious experiment by Matthew Meselson and Franklin Stahl at Caltech in 1958. They grew cultures of *Escherichia coli* in a medium that was highly enriched with the heavy isotopes carbon 13 and nitrogen 15. (The more common isotopes of carbon and nitrogen have atomic weights of only 12 and 14, respectively, so DNA built from the heavier forms is significantly more weighty.) Meselson and Stahl then transferred the *E. coli* with the heavy DNA to a medium containing only the light isotopes of carbon and nitrogen, and allowed them to divide for one generation. After one generation they found only one kind of DNA in the new *E. coli* cells, and it had a molecular weight that was intermediate between that of heavy *E. coli* DNA and normal, "light" *E. coli* DNA. So the newly formed DNA clearly consisted of one strand of heavy DNA and one strand of light. Clearly, duplication was achieved first by the splitting of the original double-stranded DNA into two single strands, followed by the replication of each one. The fact that the replication begins as soon as the splitting begins, and does not wait for a complete separation, was established later.

Next question, then: How does DNA make protein? It was clear from the start that DNA could not itself act as the manufacturer of protein. DNA is mostly confined to chromosomes within the nucleus, but it can be seen that proteins are synthesized in the cell cytoplasm, outside the nucleus. There had to be some kind of intermediary that ferried the DNA instructions to the sites of protein manufacture. This is where the second nucleic acid, RNA, enters the scene.

That RNA was in fact the intermediary between DNA in the nucleus and the sites of protein manufacture in the cytoplasm seemed likely from before the time that Watson and Crick proposed their DNA model. It was known that cells that make a lot of protein always have a lot of RNA in their cytoplasm, and the sugar-phosphate "backbones" of DNA and RNA are similar, which suggests some kind of liaison between them.

Even as early as 1953, then, the picture was envisaged: a strand of DNA, once separated from its partner, *either* can begin to make a complementary copy of itself and so replicate, *or* can begin to make a complementary strand of RNA, which then leaves the nucleus and supervises the manufacture of the appropriate proteins in the cytoplasm. The creation of RNA that is complementary to a piece of DNA is called *transcription.* In the cytoplasm, the code now carried upon this RNA is made manifest in protein by the process of *translation.*

In practice, the manufacture of protein is somewhat complicated. It involves three different kinds of RNA, each of which is created by different parts of the DNA. The RNA that ferries the message out of the nucleus is called *messenger* RNA, or mRNA. But messenger RNA does not make protein directly. Instead it cooperates with RNA that resides permanently in the cytoplasm (though it is initially manufactured in the nucleus) in bodies known as *ribosomes* and so is called *ribosomal* RNA, or rRNA. In addition, each amino acid that is lined up for incorporation into protein is chaperoned into the ribosome by a short length of RNA known as *transfer* RNA, or tRNA. Each tRNA chaperones a specific amino acid and links up to a particular place on the mRNA, with rRNA acting as a kind of workbench for this linking to take place. Enzymes, again, are needed to push this process along.

On the face of things, the DNA acts as the administrator, snugly cocooned in its office within the nucleus. The various kinds of RNA are the executors, ferrying the DNA's instructions to the protein workshops, the ribosomes, in the cytoplasm. The proteins themselves are the all-purpose functionaries, who form much of the structure of the body and make the metabolism run. Francis Crick summarized the whole process in what he called the central dogma of molecular biology: "DNA makes RNA makes protein." I like to think of DNA-RNA-protein as the Trinity—the three molecules that, by their dialogue and interaction, form the basis for all life on Earth (unless we consider that prions, the agents of scrapie and bovine spongiform encephalopathy, represent an alternative life-form). In truth, it *must* be possible for life to operate in ways other than this, for DNA, RNA, and proteins are highly evolved molecules that must have had predecessors, and those predecessors almost certainly cobbled together some form of metabolism and replication that deserves to be called living. But on Earth, at least, the Trinity has won the day, and *all* living creatures (prions and a few other oddities aside) partake of it.

So now for our third question: How does a particular piece of DNA—a gene—code for a particular protein?

One obvious and known fact was that protein consists of linear chains of amino acids (even if the final shape of a protein is far from being a linear chain) and DNA consists of linear chains of nucleotides. Somehow, the sequence of nucleotides has to determine the sequence of amino acids. This raises the problem we posed before. There are roughly 20 kinds of amino acid in protein, but there are only four kinds of nucleotide in DNA. So how do 4 code for 20?

In combinations, of course, is the answer. If nucleotides lying side by side acted in pairs, then four different kinds could produce 16 different combinations: 4×4. If they operated in threes, they could produce 64 different combinations: $4 \times 4 \times 4$. If they operated in fours, they could produce 256 different combinations:

$4 \times 4 \times 4 \times 4$. Clearly, they cannot operate in pairs, because 16 is too few. Two hundred and fifty-six is far too many. If nature is logical (and nature is sometimes logical, though this cannot be guaranteed), then the nucleotides should operate in groups of three. Sixty-four possible combinations is more than seems to be needed, but it is not absurdly too many.

Sydney Brenner and Francis Crick, at Cambridge in 1961, showed that the genetic code did indeed operate through triplets of nucleotides in DNA. They worked with mutant forms of the T4 phage, which infects *E. coli*. They found that if just one nucleotide was added to a T4 gene, the protein that resulted was nonfunctional. If two nucleotides were added, nonfunctional proteins resulted again. But if three nucleotides were added to a T4 gene, the resulting protein *was* functional. The addition of one or two nucleotides simply messed up the reading of the code, throwing the whole reading out of sync. But the addition of three nucleotides added a whole new amino acid, which changed the resulting protein somewhat, but not enough, necessarily, to disrupt its activity significantly. Each trio of nucleotides that codes for a specific amino acid is now known as a *codon*.

Brenner and Crick's work also illustrates the nature of mutation. If a gene operates—as it does—by providing successive trios of nucleotides, then a mutation could clearly be one of three kinds. If just a single nucleotide was added to the sequence, this would horribly disrupt the reading of the code. If a single nucleotide was taken away, the code would be interrupted again. But if a single nucleotide was simply changed—one substituted for another—it might or might not cause a different amino acid to be substituted in the corresponding protein, and may or may not disrupt the function of that protein. In any case, mutations in chemical terms can be tiny, but those tiny changes can have huge effects. (I am reminded of the doggerel: "For want of a nail a shoe was lost, for want of a shoe a horse was lost, for want of a horse a battle was lost." Of course, most nails lost from most horseshoes don't have such disastrous consequences. But they *can*. Similarly, alteration in, as opposed to removal of, a single amino acid may have no discernible effect at all—many mutations are neutral in their effects—but it could render the entire creature nonviable.)

Exactly which sequences of three nucleotides coded for which amino acids was established in the early 1960s, beginning with the work of Marshall Nirenberg and Heinrich Mattai, who showed that artificial mRNA consisting exclusively of chains of uracil produced a protein that contained chains of the amino acid phenylalanine. So UUU codes for phenylalanine. Similar experiments in other laboratories followed, and by June 1966 it was known that *all* codons contain three successive nucleotides, and that 61 of the 64 codons code for amino acids (some amino acids clearly have more than one corresponding codon), while the other three (UAA, UAG, and UGA) serve as punctuation marks to indicate that a particular protein should be brought to a close. In short, the genetic code was cracked.

In this way, genetics took on the additional discipline of molecular biology, and "molecular genetics" has emerged as a discipline in its own right.

This, then, is the basic fact: DNA makes RNA makes protein. But how does the collective of genes in an organism—the genome—produce a complete creature, with all its many different tissues and organs? And how does the genome provide the instructions that turn an embryo into an infant and then into an adult (which then grows old and dies)? How does the nuts and bolts mechanism—DNA makes RNA makes protein—translate into the grand life plan?

FROM NUTS AND BOLTS TO GRAND PLAN: A MATTER OF EXPRESSION

Rudolf Carl Virchow declared in 1858, *"Omnis cellula e cellula"* ("All cells derive from cells"). Thus the single-celled embryo (the zygote) divides to give rise to a multicelled embryo whose cells continue to divide until the organism reaches its final size (although some cells, like those of nerves and muscles, stop dividing before this point, while others, like those that produce the skin and stomach lining, continue to divide throughout life). This was a fine insight when the idea of the cell was still new, but of course it raises a problem. For the zygote does not divide simply to produce more cells identical to itself. As the daughter and granddaughter cells are produced, they change their character: some become liver cells, some skin cells, some muscle cells, some blood cells, and so on. This specialization of cells is called differentiation. How can one kind of cell—the zygote—give rise to cells of many different kinds?

August Weismann provided an answer in the 1880s. He suggested that as the cells divide and differentiate, they lose some of their genes. Thus the zygote contains all the genes needed to produce an entire body, but the liver cells that derive from it contain only the genes needed to make the liver, the nerve cells contain only those genes that are appropriate to nerves, and so on. Of course, in the 1880s the term *gene* had not been coined, so Weismann did not express the idea in quite this way. Nonetheless, he envisaged a steady loss of hereditary material in each kind of specialized cell line as differentiation proceeded.

Weismann was, beyond doubt, one of the greats. He was right about most things, and he should be better known to the world at large. But increasing evidence throughout the twentieth century suggested that in this particular regard he was wrong. In particular, in the 1960s the English biologist John Gurdon produced clones of frogs by the technique known as nuclear transfer. Specifically, he took nuclei from the intestinal cells of frogs, placed them in "enucleated" frog eggs (that is, eggs whose own nuclei had been removed), and grew the resultant "reconstructed embryos" into tadpoles. He thus showed that the nuclei of the intestinal cells—highly differentiated cells—*are* able to give rise to entire creatures. However, the absolute, final copper-bottomed demonstration that

Weismann was indeed wrong did not emerge until 1996, when Ian Wilmut and Keith Campbell at Roslin Institute near Edinburgh cloned Dolly the sheep. They took cells from the mammary gland of a sheep (so the cells were already differentiated), multiplied those cells in culture (which inevitably produces more differentiation), then transferred the nuclei from those cells into enucleated sheep eggs to produce a number of embryos that they transferred into surrogate mothers. One of those embryos developed to become Dolly. (In fact, two ewes produced in the previous year at Roslin—Megan and Morag—made the point just as cogently as Dolly. But Megan and Morag were made from cultured embryo cells, and the fact that those cells were already highly differentiated was not as obvious to the world as it was in the case of the mammary gland cells that produced Dolly.)

In modern vocabulary, we say that the zygote is "totipotent": that it is able to give rise to all the cells of the creature it is destined to become. In animals at least, until John Gurdon's experiments, it seemed that differentiated cells could not recover their totipotency. For three decades after Gurdon, most biologists doubted whether differentiated cells from *mammals*—as opposed to Gurdon's amphibians—could recover totipotency. But Dolly shows that if conditions are right, they can. Dolly also proves beyond all doubt that differentiated cells have *not* lost genes, as Weismann proposed. But if an intestinal cell or a mammary gland cell have not lost any of the genes that were present in the original zygote, why aren't those cells the same as the original zygote? How is differentiation possible?

The answer, of course, is that although genes are not physically lost as cells differentiate, many of them are switched off. Each cell in the body contains a complete genome, copies of all the genes characteristic of the species. But in a liver cell, all the genes are switched off except those required to produce a liver cell, and in a lung cell, only the genes needed to make lungs are "expressed." (The most obvious exception to this generalization is the red blood cells, or erythrocytes, of mammals, in which the entire nucleus, genome and all, is lost in the final stages of differentiation.) So how are genes turned on and off?

The first mechanism to be discovered was demonstrated in bacteria by the French biologists François Jacob and Jacques Monod in the late 1950s (for which they shared a Nobel Prize with André Lwoff in 1965). Specifically, they worked with *Lactobacillus,* which forms lactic acid from sugars and is involved in the fermentation of milk to make cheese. By various ingenious means they showed that each stretch of DNA that serves as a gene has a corresponding control region "upstream" of it (that is, further along the DNA molecule). Each control region has various components, including a *promoter,* a *regulatory gene,* and a *terminator.* The promoter instructs the regulatory gene to produce a protein that acts as a *repressor,* shutting down the activity of the gene it is supposed to regulate. Now it's clear that organisms such as animals and plants (collectively known as eukaryotes—that is, organisms with a distinct nucleus in each cell) have many mechanisms for controlling the expression of genes, some of which repress particular genes, and some of which turn them on. But the basic mechanism shown in *Lactobacillus* makes the general point. A gene may be present in a cell, but it remains silent and nonfunctional unless it is switched on: and there are many fine control mechanisms for ensuring that each one is turned on and off precisely to the required degree.

The general problem of development is related to that of differentiation. All creatures change as they get older: they do not merely get bigger. William Shakespeare's Jaques summarized life's progress in the human male in *As You Like It* (act 2, scene 7). He distinguished "the seven ages of man," including "the infant,/Mewling and puking in the nurse's arms"; then "the soldier, full of strange oaths"; followed by "the lean and slipper'd pantaloon," who, in the end, is "Sans teeth, sans eyes . . . sans everything."

Although the mewling infant is a very different creature in size, appearance, and behavior from the soldier full of strange oaths, the two have the same genes. It's all very strange—just as it's strange that the cells of livers and lungs have the same genes although they are so different. But again, once we grasp the notion of gene expression—genes turned on, and genes turned off—we can see how it all works. The genome can be thought of as a program. As time passes, different sets of genes within the genome are activated or switched off. Babies express genes appropriate to babies; adults express genes appropriate to adults. Behavior is affected just as much as physical appearance. To be sure, human behavior (and much animal behavior) is shaped by culture, but babies do not cry for cultural reasons, and neither does fashion account for the sexual desire of adolescents and adults (although of course it affects the ways in which that passion is expressed: whether men write sonnets or slay mammoths to impress their mates; whether women are sexually expressive or prefer defenestration to defloration).

Once we see the genome as a program, unfolding as time passes, much natural history becomes explicable that is otherwise mysterious. Why and how do caterpillars "metamorphose" into butterflies? Well, the laws of physics decree that offspring must be smaller at the time of their birth than their parents. Before those offspring can reproduce in their turn, they have to grow, or each successive generation would be smaller and smaller. So in all organisms, in general, early life is dedicated to growth, and later life—when the appropriate size is reached—is dedicated to reproduction. But growth and reproduction are very different activities, requiring very different physiological and behavioral adaptations. How can an animal change its physiological strategy as it grows bigger and older? It can switch off some genes and switch others on. You can see the changes most clearly in insects that have tight, unyielding exoskeletons like armor, which they must molt at intervals in order to grow at all. Of necessity, they progress via a series of "instars": distinct periods of growth and development between molts. They cannot simply proceed from infant to adult in a smooth continuum as soft-bodied animals may do, or as vertebrates, with their internal skeletons, can. Thus, natural selection is able to work on each instar separately, and each instar may become highly specialized. In the early life of a lepidopteran, genes are expressed that produce a voracious feeding caterpillar, which has no thoughts of sex at all. The last phase is a bright, highly mobile, winged creature dedicated entirely to reproduction. Some adult moths and butterflies (and many other insects from other classes) do not feed at all in their brief, obsessively nuptial existence.

Profound evolutionary changes can be brought about simply by changing the time switches on the genomic program: causing some genes to be expressed for longer or shorter periods; bringing others forward in the life cycle or postponing their expression. In many lineages of animals we see the evolution of neoteny: the prolonging of infant features into adulthood. We also see pedomorphosis: bringing forward the ability to reproduce into childhood. One of the most obvious examples of neoteny occurs among Mexican salamanders known as axolotls. Larval salamanders and newts are aquatic, like the tadpoles of frogs and toads, and have external gills; when they mature they emerge onto land and lose their gills. Axolotls, however, remain in water and retain their infantile gills throughout their lives. In some other salamanders neoteny seems optional: sometimes they retain their larval gills and stay in the water, and sometimes they lose them and migrate to the land. Some biologists have speculated that axolotls retain their gills because the lakes in which they live are deficient in iodine, which prevents them from producing sufficient thyroid hormone, inhibiting their maturation. But others point out that axolotls remain larval in form however much iodine they are given. More likely, they say, axolotls would find little to eat in the surrounding desert, so they have adapted to stay in the water. A small shift in the timing of the genomic program—keeping the genes switched on that produce gills—has done the trick.

Similarly, primatologists used to suggest that the flat face of human beings is a neotenic feature. Chimps have flat faces when they are young, and great prognathous (jutting) jaws when they are adult, and so, probably, did the common ancestor that chimps and humans share. Chimps retained the jutting ancestral jaw, but humans retained the flat infant face into adulthood. More detailed study of skull development suggests that this explanation will not do, however. A human skull is not molded like a baby chimp's. On the other hand, the high foreheads and wide eyes of most domestic dogs *are* a neotenic feature. Dogs resemble the puppies of ancestral wolves, in behavior as well as in looks. They have been bred that way.

Shifts in the timing of gene expression also provide a neat explanation for the otherwise mysterious process of aging—why we all, in time, become slipper'd pantaloons if the grim reaper does not call for us first. The notion here is that mutations occur all the time and that most mutations are harmful. One way for evolution to deal with harmful mutations is to suppress the expression of the mutated gene; another way is to postpone the mutated gene's expression until after reproductive age. Mutations that are postponed until after reproductive age do not get weeded out by natural selection but remain to plague the aging creature (the principle we saw with Huntington's chorea). In short, as we age, we run into a backlog of postponed mutations. You can run, as they say in the westerns, but you can't hide.

The general picture of life's controls that emerged as the twentieth century progressed is wonderfully neat. Archibald Garrod's "one gene, one enzyme" became Francis Crick's "central dogma": DNA makes RNA makes protein, and proteins, of course (mainly in the form of enzymes), are the body's functionaries, effectively running the show. The sequence of nucleotides in the DNA corresponds to and codes for the sequence of amino acids in the protein; and the sequence of amino acids determines how the finished protein will fold into its final three-dimensional form, and hence how it will behave. Once we see how genes can be switched on and off, the phenomenon of differentiation ceases to be a mystery and we can see how the whole genomic program unfolds to produce creatures that change as they proceed from embryo to adult to geriatric, a program seen in dramatic form in the caterpillar's progress to the butterfly, and in all of us as we journey from mewling infant to slipper'd pantaloon.

The experiments outlined in this chapter illustrate the general nature of science. Huge problems—like what is a gene made of?—are tested by irreducibly simple experiments: for example, the demonstration that bacteria cannot transmit hereditary information if first treated with an enzyme that destroys DNA. The experiments may be technically difficult—it really is hard to manipulate individual cells—but they are simple in concept. Then, from a minimal baseline of "robust" observations, scientists *infer* how life must be. Nowadays, with ultramodern microscopes, it is possible to photograph DNA molecules in fine detail, and to show that they are indeed as Crick and Watson described. But their initial description was based on inference, drawn from a wide range of disparate data. *All* science, when you boil it down, is inference, but it works nonetheless. This is what scientists mean by *elegance:* to build great robust conceptions of the world on minimalist observations. Mendel, with his rows of peas, showed the principle supremely.

However, the picture revealed in the early and middle decades of the twentieth century was a little too neat. Biology is rarely neat, at least in all its details. We and our fellow creatures were not, apparently, designed to a specification, in the way that Ferrari designs cars and Sony designs televisions. Nature evolved, and systems that evolve are messy. Each generation builds on what was there before, without necessarily disposing of its predecessors, so that every living creature—and every system within every creature—is liable to carry legacies from its ancestors, in the form of "vestiges." Underlying the neatness, then, and to some extent subverting it, we should expect to find all kinds of diversions, curlicues, accidentals, and conditional clauses, and indeed we might find that the true nature of the beast is not as we initially envisaged it at all. As the twentieth century wore on, the initial simplicities became a little more, well, *lifelike.*

LIFE IS FULL OF SURPRISES

An early surprise was that DNA—including functional genes— occurs in the cytoplasm of the eukaryotic cell as well as in the nucleus. Mitochondria, the small cytoplasmic structures ("organelles") that contain the enzymes associated with respiration, also contain their own genes. So do chloroplasts, the organelles in plants that carry the green photosynthetic pigment, chlorophyll. Mitochondrial genes account for only a few percent of the cell's total genes, but they are influential nonetheless. They provide at least some of the information needed to make the respiratory enzymes. Mutations in the mitochondrial genes can lead to definable defects. The presence of mitochondrial DNA supports the idea, championed by the American biologist Lynn Margulis, that the eukaryotic cell is a coalition. The theory is that the cytoplasm of the eukaryotic cell came from the bodies of ancient prokaryotes of the kind known as archaea, and that the mitochondria evolved from parasitic bacteria that lodged within the body of the archae, and stayed. Host and lodger then coevolved, and many of the invader's genes passed to the nucleus of the host (so that mitochondria are no longer able to live independently, outside the cell). But the mitochondria have retained some of their ancient bacterial genes. In practice it is clear that mitochondria do resemble some living bacteria—notably alpha-proteobacteria—and their DNA is in some ways more like that of a bacterium than that of a eukaryote.

It also came as a shock to discover from the mid–twentieth century onward that DNA itself is much more untidy than was anticipated. Molecular biologists might have expected to find that each double helix of DNA consisted entirely of genes, strung without interruption from one end to the other, each coding for proteins, tRNA, or rRNA, with occasional codons to provide punctuation and stretches to regulate expression. Not a bit of it. Between the genes there are huge stretches of DNA that do not code for anything at all. Worse, there are similar stretches *within* the genes, known as introns (while the bits of the gene that do actually code for proteins are known somewhat confusingly as exons). Originally, the noncoding DNA was known as junk DNA—although it seems presumptuous to suggest that something is "junk" simply because its function is unknown. Clearly, at least some of the alleged junk does serve some function. Introns, for example, may be involved in gene expression. Be that as it may, about 90 percent of the DNA in a eukaryotic cell is now known to be noncoding. The DNA of bacteria is much tidier, with very little noncoding DNA indeed.

The noncoding DNA evidently has various origins. Some is clearly vestigial: old genes that have lost their function but have not been weeded out; old viruses that have become integrated into the host's own DNA. After all, if the noncoding DNA is not expressed, then it does the creature no harm, unlike some deleterious alleles. If it does no harm, then it is not visible to the forces of natural selection and remains to clutter up the genome. When genes are transcribed to make mRNA, the introns are slavishly copied as well but are snipped out when the mRNA exits the nucleus, so that when the mRNA reaches the ribosomes, it is strictly functional.

One of the greatest of all surprises was revealed in the 1940s— before the structure of DNA was known in detail—by the American geneticist Barbara McClintock. She showed that some genes actually shift their position within the genome. She called them mobile elements, but they are also known colloquially as jumping genes, some of which are known more formally as transposons. McClintock first revealed such mobile elements in corn, but they are known to be ubiquitous.

In fact, Barbara McClintock's mobile elements helped promote a new way of thinking about DNA that has had profound theoretical consequences. Biologists have naturally tended to think—as Darwin did—that the "unit" of life is the individual: the particular cat or oak tree; the particular human being. If you think of individuals as the basic units, then you assume that the genes, which help shape the individual, are somehow subservient to that individual. They are parts of the individual, just as carburetors and brake shoes are parts of cars. You expect the carburetor and the brake shoe and all the other component parts to do what is good for the car. But there is something anarchic about a jumping gene, as if bits of DNA were simply doing their own thing.

Indeed, there is a great deal about the jumping gene that is anarchic. A stretch of DNA detaches itself from its allotted place in a chromosome and then inserts itself somewhere else, and where it ends up seems to be more or less random. A jumping gene may indeed cause mutations in the places where it lands. If it lands in the coding region of the gene (in an exon, rather than an intron), then the mutation it causes could be damaging. In fact, some biologists have been so impressed by the extent of such activity within the genome that they suggest it may be a more important agent of change than natural selection itself. Gabriel Dover of Cambridge University coined the expression "molecular drive"; he suggests that molecular drive, operating at the level of the DNA itself, has largely determined the way we are. Others acknowledge that molecular drive is a fact—that DNA is far more restless and apparently anarchic than might be supposed—but in the end, the DNA itself survives only if it manages to produce a creature that can actually function. If the restless DNA produced creatures that could not compete with their fellows, they would die out, and their unruly DNA would die with them. In short, the mechanisms of molecular drive may cause variations in lineages of creatures, but in the end the different variants are weeded out by natural selection. So, molecular drive exists and is important, but as an agent of evolutionary change it seems to offer very little challenge to natural selection.

Allied to this realization—that individual bits of DNA may behave apparently anarchically—is the concept of "selfish DNA." The selfish DNA idea is not quite the same as the selfish gene idea, which is discussed in chapter 6. The concept of the selfish gene encapsulates the notion that natural selection operates most forcefully at the level of the gene rather than of the individual as a whole. As we will see, this notion is much misunderstood and contains the apparent paradox that *because* genes are selfish, the creatures that contain them can behave most unselfishly, and indeed altruistically, to the extent of self-sacrifice. Thus the concept of the selfish gene belongs to evolutionary theory as a whole. The concept of selfish DNA belongs to molecular biology. It describes the odd but inescapable fact that individual bits of DNA do indeed do their own thing. If these bits of DNA had any self-awareness, they would see themselves *not* as servants of the organism of which they are a part but as individuals, as free citizens milling about in the marketplace.

One highly intriguing manifestation of "selfish" DNA is the phenomenon of "genomic imprinting." You will recall that Mendel was bedeviled by spermists and ovists, who had been so slyly lampooned by Laurence Sterne in the previous century. He took great pains to show that "factors" inherited from the female line are entirely equivalent to those inherited from the male line, and announced the fact triumphally in his paper of 1866. From the 1980s onward, however, it has become clear that he was not entirely correct in this. While this generalization applies to most creatures, in mammals the action of at least some of the genes *is* influenced by the sex of the parent.

In 1980 scientists at Cambridge, notably Azim Surani, found that it was impossible to make a viable mouse embryo from two pronuclei that were both derived from sperm or both derived from eggs. Embryos made with two sperm nuclei developed excellent, lusty placentas, but the body of the embryo itself was feeble and nonviable. Embryos made with two egg nuclei did well at first, but they failed to develop placentas that were able to sustain them. Nuclei from both egg and sperm were needed. Thus parthenogenesis—virgin birth—seemed to be impossible in mammals, at least by this simple route, since a male nucleus was needed. More broadly, it was clear that male and female nuclei were not exactly functionally equivalent. A sperm nucleus might contain the same genome as the egg (give or take a few allelic differences), but at least some of the genes derived from the male were behaving differently than would their homologues derived from the female.

This is genomic imprinting. In mammals at least, the male (or the female) does indeed impose his (or her) special stamp on at least some of the genes. The defective allele that causes Huntington's disease is imprinted in just this way. If the allele is inherited from the father, the disease strikes earlier, and in general is worse, than if the allele is inherited from the mother. A daughter of an affected father may suffer badly, because she has inherited the bad gene through the male line, but she then puts her own stamp on the allele, and her own offspring will generally not suffer the disease until they are older.

Huntington's is a pathology—it may be seen as a bolt from the blue—but we must ask *why* the phenomenon of imprinting should have arisen. The question is answerable only in terms of selfish DNA. In mammals, the fetus can be seen (physiologically speaking) as a parasite. A gene that causes the parasitic fetus to take more than its proper share of nutrient will produce a big, lusty infant but also stands a good chance of killing the mother. Analysis reveals that genes passed on through the male line will spread through the population if they do indeed produce good, healthy infants, even if they do harm the mothers that bear them along the way, while genes passed on through the female line will do less well if they cause harm to the mothers along the way. Hence, broadly speaking, some male-derived genes have evolved to produce greedy fetuses, while female-derived genes produce less voracious fetuses. In any one fetus, alleles derived from the male parent compete with alleles derived from the female parent. The picture is further complicated because both sexes of fetus, of course, contain genes derived from both sexes of parent. There are female-imprinted genes in little boys, and male-imprinted genes in little girls. All in all, then, the theory of genomic imprinting requires a cool head. The generalizations are as stated, however: each sex of parent imposes its own stamp on some of the genes; and the phenomenon is explicable only when we consider genes as selfish entities.

How, then, does selfish DNA—genes battling it out with each other within the cell—manage to produce a genome that in the end provides the code for a beautifully coordinated mushroom, horse, or human being? The answer may be essentially the one by which Adam Smith, that eighteenth-century paragon of the Scottish Enlightenment, explained the way that cantankerous individual human beings manage, between them, to produce coherent societies. Each individual, said Smith, pulls on his or her own rope. But human beings do not thrive well on their own. They do better as members of society. Ergo, individuals who are acting entirely in their own interests thrive best by doing whatever meshes in with what others are doing. Through all the individual acts of selfishness, then, what Smith called the hidden hand produces a collection of human beings who to all intents and purposes are cooperating. The selfish individuals form a society because it is in their selfish interests to do so. As Richard Dawkins has said, individual rowers do best when they cooperate with seven others to produce a rowing eight. Eight self-centered rowers fired by personal ambition can produce an eight that functions just as well as one in which each individual thinks only of his team's glory.

But it is also clear, as Adam Smith pointed out, that human beings (and other animals) *also* have a sense of "society as a whole" and work to make a better society. Even without this, though, Smith's concept of the "hidden hand" seems to explain well enough how blind and selfish genes end up cooperating.

Yet in recent years, many a liberal and socialist has opposed the notions of the selfish gene and selfish DNA on political grounds. Some biologists have allowed their politics to get in the way of their scientific detachment and have attacked biological findings that they ought to realize are actually rather strong. We might simply point out that science and politics should be kept separate. The task of science is to tell us, as far as can be found out, what is actually true, and the task of humanity, as moral creatures, is to decide what kind of societies we want to create and to set out to do what we think is right whatever science tells us is actually the case. But it is tempting to point out that socialist objections to selfish genes and selfish DNA are simply muddleheaded. These are the ideas that have explained Darwin's dilemma: that animals often, and in very many contexts, do behave unselfishly and even self-sacrificially. The selfish DNA/selfish gene concepts show, in fact, that nature is not "red in tooth and claw," as Tennyson supposed. Of course all creatures—and all genes—must compete with one another: that is life's burden. But in reality, creatures (and individual genes) often survive best by cooperating. Life itself evolved as a coalition; it could not have evolved in any other way. In short, if biological theory showed that we are pressured to be at each other's throats, then we should learn to live with that fact and attempt to override it. Fortunately, however, biological theory shows no such thing. Darwin never said it did.

In fact, it is time to look at what Darwin did say, and at how his ideas jibe with those of Gregor Mendel. It took biologists a few decades to perceive that Darwin's evolution and Mendel's genetics do indeed fit together; once they did see this, the result, known as neo-Darwinism, became one of the great intellectual insights of the twentieth century. The synthesis of the two sets of ideas is the subject of the next chapter.

Mendel and Darwin:

Neo-Darwinism and

the Selfish Gene

Mendel (and others in the German tradition) provided half the foundation of modern biology. Charles Darwin, in England, provided the other half. For Darwin produced the first truly plausible theory of evolution: "by means of natural selection," an idea that came into the world with the power of revelation. Natural selection may yet prove to be the most fundamental "force" at work in all the universe. Even physicists, these days, wonder whether their "fundamental" particles are really so fundamental after all, or whether they too, like elephants and oak trees and human beings, have been selected from among a range of candidate particles.

Yet Darwin—like Mendel—is easy to underestimate. People seem to misconstrue his genius, how great he was, perhaps because he seems so homely (tweedy, rich, amiable, established middle-class, puttering around his garden in Kent, surrounded by children and playing backgammon with his wife) and because he apparently deals in the day to day. *On the Origin of Species* really was seminal—modern theoretical biology really is an addendum with one significant codicil and many footnotes—but it tells of pigeons, orchids, ants, and barnacles. Anyone can respond to such exemplars: our brains are geared to natural history. Really clever people, we tend to think, deal in weirdness that most of us will never grasp: Einstein with relativity; Niels Bohr with quantum mechanics and the insubstantiality of the Universe. Metaphysically speaking, though, and for that matter in sheer brain power, Darwin was among the greats, perhaps, we might whimsically suggest, playing Tolstoy to Einstein's James Joyce. Don't be fooled, either, by his puttering; he just happened to be, by way of a hobby, the finest field naturalist of all time, with a canny line in experiment. Of course, he makes use of other people's observations in *Origin* (and gives them fulsome credit), but it's astounding how many insights are his own, from ants in Hampshire and the "tangled banks" of Kent to the entirety of South America.

Then again, many of those who continue to underestimate Darwin simply misunderstand history. It is often said that when Darwin wrote *Origin,* evolution was "in the air." Darwin himself acknowledges many of his predecessors in his introduction to *Origin.* Jean-Baptiste Lamarck was among several biologists in the late eighteenth century and early nineteenth century in France who expounded evolutionary ideas. We have seen that in Moravia, Mendel's teacher Unger wrote in an evolutionary vein, and so did Mendel, during one of his unsuccessful attempts to qualify as a teacher. In England, Darwin's grandfather, Erasmus, wrote in the eighteenth century of creatures emerging from earlier creatures. In Scotland, in the 1830s, Patrick Matthew summarized the gist of natural selection somewhat bizarrely in the appendix to a book on naval architecture.

But none of these apparent predecessors hit the jackpot. Many, like Erasmus Darwin, felt that living things had changed over the years but proposed no mechanism; in particular, they could not explain how creatures are clearly adapted to the places where they live: shaggy where it's cold, and smooth where it's hot. Lamarck proposed a mechanism—"inheritance of acquired characteristics"—but it failed to convince at the time and is now known to be wrong. German biologists, like Unger, tended to conflate evolutionary change from generation to generation with the development of the embryo. Matthew and others who did think in terms of natural selection failed to develop their ideas. Perhaps they simply did not perceive, as Darwin did, that to produce a theory of evolution that could be applied universally, and was robust, it was necessary to look at the natural history of the whole world. It was also necessary to take account of paleontology (fossils played a large part in Darwin's thinking), embryology, and the homely craft of livestock breeding (which employs the "artificial selection" that gave Darwin the clue to "natural selection"). Darwin—like Matthew—conceived of natural selection in the 1830s, but unlike Matthew, he realized that a single flash was not enough. Darwin knew that he could not make the case that was needed unless he brought to bear a huge weight of evidence and thought through the minutest caveats, which is why it took him 20 years to publish. The only other biologist to grasp the full idea was Alfred Russell Wallace, who summarized the notion convincingly and succinctly in 1858. But Darwin is rightly acknowledged as the greater.

When Darwin wrote *Origin,* the dominant belief among educated people—and of course, among the uneducated—was "special creation." God had simply made the creatures that we see around us in their present form and placed them in their present locations. Adaptation was not perceived as a problem; of course God had matched the creatures to their environments. Neither was there any crude division of belief between churchmen and scientists. Most of the influential scientists of the day were creationists, including the great Adam Sedgwick, who was Darwin's teacher, and the indisputably great Richard Owen, who, in the early nineteenth century, was Britain's leading biologist. Owen coined the word *dinosaur,* and his ideas in anatomy are still cited. Even those who accepted some idea of evolution tended to reject the notion that any one lineage could "branch." At best, they envisaged that God had created a range of primordial creatures some time in the past (and the world was commonly perceived to be only about 6,000 years old) that had then all "evolved" quite independently, up a series of separate evolutionary ladders.

So, as Ernst Mayr describes in *Towards a Philosophy of Biology,* Darwin needed to undertake three conceptually separate tasks. First, he needed to show that evolution was a fact—that it was the route by which all existing creatures had assumed their present form, which emphatically was not the common view. Second, he had to provide a plausible mechanism to explain how change over time led to adaptation, and this he did by proposing the mechanism of natural selection. Third, he had to establish what to us now seems obvious but in the nineteenth century decidedly was not: that species can change. "Species," in the mid–nineteenth century, was still essentially a Platonic concept; even biologists still felt that the creatures they saw about them were mere material copies of some heavenly ideal. It was inconceivable that they could change into different creatures, or that a lineage could branch, with any one species giving rise to many different ones. Indeed, "transmutation of species" was virtually a blasphemy. As Mayr says, Darwin's principal battle was not with God but with Plato. There cannot be branching of lineages without transmutation of species; and once there is branching of lineages, it is possible, as Darwin made explicit in *Origin,* to envisage all creatures united in one great universal tree of life that sprang from a single, common ancestor. Thus eagles and oak trees are our relatives, and, in fact, it now transpires, even the oaks are not so distant from us as it might be supposed. Many other creatures, including the homely amoeba that every biology student draws at some time, are far more distantly related to us than plants are.

Natural selection was the notion that carried the day, the mechanism that made the idea of evolution plausible. The bones of the idea, as with all the greatest ideas, can be stated in a couple of paragraphs. First, all creatures are engaged willy-nilly in what Darwin called a struggle for existence. The English economist

Thomas Malthus had pointed out at the end of the eighteenth century that if the human population continued to grow as it was then doing, it would soon exceed resources, and Darwin saw that this principle applied to all species. Every creature could, if unconstrained, produce enough offspring to cover the entire Earth within a few hundred years, and the fact that they do not do so shows that they are indeed held back. Many more are born than live to reproduce themselves.

Second, creatures that are of the same general type nonetheless vary. No two kittens in a litter are exactly the same. Inevitably, some of the variants will be better adapted to the prevailing conditions than others. The better adapted will be more likely to survive and reproduce. Herbert Spencer coined the expression "survival of the fittest," where "fit" has the Victorian sense of "apt," and Darwin later adopted this form of words. The survivors' offspring will resemble themselves because "like begets like," although the resemblance is not exact for all the reasons we have already outlined in this book. Hence, over time, lineages of creatures become better and better adapted to their surroundings.

Darwin did not stop with natural selection, however. Indeed, it has been said that after *Origin,* he seemed to lose interest in it. After all, he had been living with the idea for some decades. He went on to consider another possible mechanism of evolutionary change—sexual selection—which he presented in 1871 in *The Descent of Man and Selection in Relation to Sex.* Here he emphasized that sexually reproducing creatures (including most animals and plants) cannot reproduce unless they find compatible mates, and that, most obviously in the case of sentient and mobile animals, they first have to attract mates. Therefore, he argued, selection will favor individuals that potential mates find most attractive. Hence, for example, the peacock's tail.

Often, the features that are most attractive to mates detract from a creature's ability to withstand the rough-and-tumble of everyday life: the peacock's tail does not do much for his agility.

Yet sexual selection may also encourage the evolution of characters that are of enormous value in other contexts; Geoffrey Miller of University College London argues that only sexual selection can explain the extraordinarily rapid evolution of the human brain (see his book *The Mating Mind,* William Heinemann, London, 2000). All in all, biologists are only now beginning to appreciate the full significance of sexual selection. But Darwin saw it first, and it took the rest of the world more than 100 years to catch up.

Darwin, his own best critic, perceived the two most important flaws in his own evolutionary thesis. First, he had no plausible mechanism of heredity, nothing to explain convincingly how "like begets like" and yet gives rise to variations. Second, his ideas were unquantified. He was aware of his mathematical deficiency at a very early age. In 1828, at the age of 19, just after he had been sent down from Edinburgh University, he wrote to a friend: "I am as idle as can be: one of the causes you have hit on, viz irresolution, the other is being made fully aware that my noddle is not capacious enough to retain or comprehend Mathematics. Beetle hunting and such things, I grieve to say, is my proper sphere." (Actually, the list of Edinburgh alumni who failed to finish the course is almost as long and distinguished as of those who did.) Math is crucial in the development of science, of course, as Mendel showed. But science also needs creative imagination, and Darwin showed that those who have an abundance of it may make pivotal contributions even without math. To be sure, very few great scientists were nonmathematical, but Darwin was one of them. Michael Faraday, intriguingly, was another. But in the 140 years since *Origin,* the deficits that Darwin perceived in his own work have been made good. In that 140 years, too, biologists have made just one, albeit enormously significant, adjustment.

The first of the great problems—the mystery of heredity—was, of course, effectively solved within a few years of *Origin* by Mendel. But we have seen that although Mendel had many good friends, he did not seem able to gain their serious attention. Darwin's experience was otherwise. He leaned heavily on the great geologist Charles Lyell, who was older than himself, and on his juniors, Thomas Henry Huxley, Joseph Hooker, John Lubbock, and others, who assumed the role of critical disciples, supportive but sharp in equal measure. Contrast Mendel's dismal correspondence with Karl Naegeli. So it was that Darwinian evolution emerged as the great obsession of the late nineteenth century while Mendel's genetics was left to become the science of the twentieth century.

You will surely see intuitively that Mendel's genetics and Darwin's ideas on evolution fit together beautifully. Strangely, however, when Mendel's ideas on heredity were first rediscovered, they were felt to be *at odds* with Darwin's theory. The history of science is the history of human beings; it is not the logical outworking that nonscientists and bad philosophers of science suppose it to be. The problem was (or so it was perceived) that Mendel's rules were based on patterns of inheritance of simple characters that were coded by single genes. If all characters were of this kind, then evolution would surely follow a very jerky course. Any one character would have, as Mendel said, an "all or none" quality. But Darwin had stressed that the characters of all creatures tend to change gradually, step-by-step, over many millions of years.

The answer to this apparent contradiction now seems obvious, and would have been obvious to Mendel, if he had lived long enough to be consulted. Mendel deliberately set out to study very simple, single-gene characters because, as he well knew, these alone would give clear results that could be analyzed statistically and reveal the underlying patterns, and hence suggest the underlying mechanisms. But he also knew full well that most characters in most creatures do not have an all-or-none quality. Most, as he certainly realized, are polygenic. A character that is polygenic can obviously be changed gradually as the generations pass simply by changing one gene at a time. So there is no irresoluble conflict. Indeed, there is no conflict at all.

All this mercifully became obvious during the first decades of the twentieth century, and by the 1940s Mendel's genetics was fully reconciled with Darwin's idea of evolution by natural selection. The reconciliation is known as the modern synthesis, and the result of this synthesis is called neo-Darwinism. Neo-Darwinism is the cornerstone of modern biology and can properly be seen as one of the intellectual triumphs of the twentieth century. Stir in molecular biology, which came on-line after the 1950s, and you effectively have a summary of modern biology. In practice, however, the true significance of the modern synthesis seems underappreciated, at least by nonbiologists. Physicists are more impressed by quantum theory, and nonscientists are more impressed by physicists.

Neo-Darwinists essentially perceive that natural selection affects the gene pool as a whole. As time passes, duplication of genes and genetic mutation tend to increase the total variety of alleles within the gene pool, and hence the total variation of the creatures themselves. Natural selection then weeds out the individuals that contain the least favorable combinations of alleles. As time passes, some alleles are totally weeded out. Thus, as the generations pass, the composition of the gene pool—the frequency of the different alleles within it—changes. As we will see, late twentieth-century neo-Darwinists tend to argue that although natural selection in practice appears to operate most directly on individuals—it's the individual creatures who die, or live to reproduce—it is better, in general, to think of natural selection as acting on individual genes. The general point is that the action of natural selection alters the composition of the gene pool over time.

Darwin's second great desideratum—the need to manipulate his evolutionary notions with the precision of math—began to be tackled seriously in the 1920s and 1930s by the great British biologist-mathematicians R. A. Fisher and J.B.S. Haldane. But the math that has proved truly appropriate was developed only in the

1930s, by the Hungarian John von Neumann: the notion of "game theory." Von Neumann himself did not apply game theory to problems of biological evolution; that was done by the British biologist John Maynard Smith. Maynard Smith in particular showed how with game theory it is possible to measure and compare the effects of different survival strategies to measure aspects of "fitness." Although Darwin was embarrassed by his own poor numeracy and impressed by mathematicians such as his cousin Francis Galton, he felt deep down that the math of his day was simply not able to tackle the kinds of problems he was posing. In practice, in the absence of game theory, his ideas were hijacked by dull statisticians like Galton and Pearson, with their bell curves of normal distribution and the rest—a line of thinking that has led many biologists and sociologists to emphasize the differences between human beings, and has helped give Darwin a bad name in some circles. In fact, all those decades of grind can be seen as a diversion, not to say an aberration. I am sure Darwin knew in his bones that this was not the way to go.

These two twentieth-century innovations—genetics and the appropriate math—can be seen as a fleshing out of Darwin (although of course both Mendel and von Neumann must be acknowledged as great in their own right). The third twentieth-century innovation is truly an adjustment. It concerns the level at which natural and sexual selection operate. Darwin supposed that selection operated on individuals: the cheetah that runs more quickly than other cheetahs is favored; the peacock with the largest tail catches the hens. But since the late twentieth century, Darwinians—first the avant-garde, then the mainstream—have thought of selection as acting on individual genes. In the present simplistic examples, a gene that promotes swiftness is favored in the cheetah, a gene for attractiveness is favored among peacocks. This is the concept that Richard Dawkins has encapsulated as the "selfish gene." Whole organisms, he says, are merely vehicles for genes. Often there is no practical reason to distinguish between selection at the level of the individual and selection at the level of the gene. After all, a gene that makes a cheetah swift also helps the whole animal to survive and reproduce; and if the gene fails, it's the whole animal that dies without issue. But whereas thinking at the level of the individual works some of the time, thinking at the level of the gene works all the time; it is like the difference between Newtonian mechanics and Einstein's relativity.

The expression *the selfish gene* is wonderfully arresting. But when Richard Dawkins made it the title of his brilliant book of 1976, he did not anticipate that so many people would read the cover but skip, ignore, or otherwise misconstrue the text that followed. On the face of things, after all, the idea of the selfish gene seems to imply that the possessor of such a gene must itself be selfish. Since all genes are supposed to be selfish, this implies that all creatures are selfish, including all human beings. This would be a dismal thought indeed, and this is what many people evidently think Dawkins intended to imply.

But actually, as Dawkins is at pains to explain, the complete opposite is true. Darwin wondered, in *Origin,* how it was that ants could cooperate the way they do, apparently sacrificing themselves for the good of their peers. This apparent "altruism" did not seem to square with the notion that creatures are condemned to compete with one another. But in the 1960s and later, Bill Hamilton, who became a professor in the zoology department of Oxford, solved Darwin's dilemma by reference to the concept of what Dawkins called the selfish gene. The point is that *because* the gene is "selfish," it is "concerned" only with its own replication; and a gene that sacrifices its owner in some act of altruistic foolhardiness will nonetheless be disseminated if, through the apparent self-sacrifice, it thereby enhances the survival of other individuals who contain copies of itself. Hence animals—including humans—will sacrifice themselves for their own relatives, and especially for their own children, whom they know are likely to carry copies of their own genes. This is "kin selection": simple in principle, but once you start to apply it, remarkably powerful.

In fact we cannot really see the full power of Darwin's idea of natural selection until we apply it directly to genes. For the word *selfish* is merely meant to be an intriguing slogan, a flag to entice people to look at the ideas. It does not literally mean "selfish" of course, because "selfishness" implies some underlying emotion, and genes are simply bits of DNA when you boil them down, with no emotions. This is the point: they are without emotion and without senses. Their role in life, their raison d'être, is simply to replicate. Those that replicate remain in the population and spread, from generation to generation; those that do not, die out. That, at bottom, is all there is to it. It is a circularity: nothing succeeds like success; only success succeeds. It is *meant* to be a circularity. If you perceive the circularity of natural selection, you begin to feel the weight of this relentless force, which certainly shapes living creatures and possibly shapes the whole Universe, down to what are thought of as fundamental particles, and even the apparently immutable laws of physics.

Darwin had two other great and pervasive ideas. The first was that behavior could be and is shaped by natural and sexual selection just as physical features are. A peacock's tail is simply a waste of energy if the peacock does not strut his stuff, and if the hen does not respond to his strutting. The cheetah's beautiful shape and extraordinary physiology are mere caprices if the cheetah does not care to give chase. Behavior must evolve too.

Darwin's final great and revelatory idea—which offended many of his contemporaries and successors even more than the idea of evolution itself—is that there is continuity between human beings and other creatures. Everyone knows that he proposed that human beings have indeed evolved, just as other creatures evolved. The story of Adam and Eve has been, in its literal form, another casualty. Everyone knows, more particularly, that he proposed in *The Descent of Man* that human beings are most closely related to chimpanzees and gorillas, and that we and they share a recent common ancestor who lived in Africa. The weight of evidence since then suggests that in all these details he was absolutely right, just as he was uncannily right in almost everything he looked at (from sex in barnacles to the labors of earthworms to the propensity of some flowers to track the path of the sun).

Put these two thoughts together—that behavior has evolved by natural selection, and that human beings have evolved, just like every other creature—and it becomes reasonable to contemplate the evolution of human behavior. There have been several false starts in this area, notably by Herbert Spencer in the late nineteenth century, who first, and prematurely, devised the notion of "social Darwinism." But biologists of the late twentieth century, informed by the notion of the selfish gene and restrained by the discipline of game theory, have been able to make a much more orderly and convincing attack. The moderns, in short, produced the discipline of sociobiology, out of which emerged evolutionary psychology. We will see where this is leading in the next chapter.

SEVEN

Genes for Behavior:

Evolutionary Psychology and

the Nature of Human Nature

On an ancient wall in Delphi is inscribed the injunction "Know thyself." It does not follow that if we know ourselves we will necessarily behave better, yet self-knowledge should surely enable us to behave more appropriately. On balance, it seems to be a good thing. Evolutionary psychology is an attempt to obey that Delphic command: it is an attempt to explore our own minds, and thoughts, and behavior, to ask *why* we feel, think, and behave as we do. It is, in fact, an attempt to pin down that elusive quality that we refer to casually as human nature.

Of course, *all* psychology aspires to do at least some of this. But the specific approach of evolutionary psychology is to apply *evolutionary* ideas to what we know about our own thoughts and social interactions, to see what light might be shed. Evolutionary psychologists do not for the most part seek to sweep aside all other kinds of insight into the human psyche—from other fields of psychology, from sociology, or indeed from literature. But they do aim to put the insights of evolutionary biology into the ongoing discussions, in the hope that understanding will be nudged along. Evolutionary psychology is a young subject—so far, there are not many extensive, robust studies—but already the signs are that it has a great deal to offer. Some of its ideas do not always seem to be politically correct, prompting some of its critics to protest far too loudly, sometimes with some cogency, but usually without revealing anything that would qualify as understanding. When the nonsense has died down, evolutionary psychology will surely prove to be one of the most significant intellectual and moral legacies of the late twentieth century. Most of the chapters of this book could be expanded into fat books of their own, but this one more than most. In fact, though, this account will have to be a telegraphic run-through of the main ideas—not exploring them in depth, alas, but trying to show how the different notions fit with each other, and with ideas from other fields.

Is there really such a thing as human nature? The idea was certainly fashionable in the eighteenth century; it was a key theme of the Enlightenment. David Hume published his *Treatise of Human Nature* in 1739 and 1740, Alexander Pope wrote that "the proper study of mankind is man," and Samuel Johnson, Rousseau, and Voltaire all offered accounts of what human beings are really *like.* But others, notably John Locke in the seventeenth century, argued that the human mind is, as he said, a tabula rasa, a blank tablet; and many have taken this to mean that human beings can be shaped absolutely by their environment. Bring a girl up as an Irish Catholic and she may become a nun; raise the same young woman as a Soviet atheist and she will raze the convents to the ground. If Locke's idea is taken literally, then we are all protean creatures, able to be molded every which way, and it hardly seems reasonable to speak of human nature at all. We simply become as life shapes us, unless, of course, we say that it is "in our nature" to be infinitely flexible.

Actually, nobody (I presume not even Locke) would care to argue that human beings are *infinitely* flexible. There are many tales in mythology of human babies raised by animals, from Romulus and Remus, the founders of Rome, who were nurtured by a wolf, to Edgar Rice Burroughs's Tarzan, brought up by apes. All learned some of the tricks of their adoptive parents' societies, but all broke away. This is fiction, but surely it reflects what is the case. If you were brought up by giraffes, you would doubtless learn to browse from high places, but you would always feel—would you not?—that this was not quite your métier. When tourists appeared in the savanna you would at least be intrigued, just as Tarzan was when people finally came his way. You would certainly not turn away as disgustedly as the giraffes would. You would want to say to the chief giraffe, "Hang on a minute! Let's see what's going on here!"

In such discussions we may choose to refer to texts and authorities (if such there are), but it is hard to improve on common sense and observation. Common sense and observation should at least provide the initial hypotheses. They tell us that we are flexible creatures, both physically and, much more importantly, behaviorally. The same person born in different places and circumstances really might develop in a variety of ways: to behave in the accepted manner of a laborer or aristocrat, midwesterner or Bostonian, faithful husband or polygamous despot, pillar of society or Mr. Big, shrinking violet or charismatic leader. Great actors show how easily, in principle, a single individual may slip from role to role. Literature is full of examples, from "Cinderella" to Mark Twain's *The Prince and the Pauper.*

Yet we *know* that however many "roles" we may adopt, however our talents and even our personalities may be molded by our experience, we remain recognizably human. A Martian, taking a long view of humanity—as it is difficult for humanity itself—would surely be much more impressed by the similarities between us all than by the differences. Some qualities that human beings have in common with each other they also share with other species. For example, humans are obsessive child rearers, but then so are virtually all mammals and birds. Sometimes the mother shoulders the burden alone, but among humans the father generally shares at least some of the work, and this is true too of many birds. Other qualities are uniquely human, including language, of the kind that is underpinned by syntax and is, apparently, open-ended and infinitely flexible. Of course, different people speak different languages, and the range of accents and dialects within any one language can be truly staggering. It used to be possible in Britain (and perhaps still is) to trace people's origins to their particular villages, and to track all the changes in social fortune that life has imposed on them, just by listening to their voices. The flexibility—the subtleties of ear and larynx, and the processing of information that goes along with them—truly surpasses understanding.

Even more striking, however, is that all those different people share the ability to learn their own language, with all its quirks of accent and dialogue, within the first few years of life, given only a minimum of always rather scrappy clues. *All* human beings can do this, in the absence of gross pathology. No other creature can. What we all share, too, is an almost total inability to acquire a second language, later in life, to the same degree of surety that we once learned our own. We may speak a foreign language correctly, but the languages that are truly ours, we can abuse and reinvent on the hoof, which is a much more cogent skill.

So yes, human beings are supremely versatile creatures, much more so than giraffes, who merely browse from the tops of trees, or even wolves or apes, although their lives are highly various and differ in tradition from time to time and place to place. Yet all human beings are recognizably human beings and not something else. Different individuals and different societies do things in different ways, but at a not-particularly-deep level they all do the same kinds of things, some of which some other animals do as well, and some of which are uniquely human. It seems reasonable to call the things that all human beings do human universals, and to suggest that the sum of human universals together forms this elusive quality of human nature.

But of course, *pace* Locke (although this is really just another commonsense observation), the concept of human nature, defined by universals, does not imply for one second that each individual or each society is *bound* to do particular things in particular ways. The point is merely that although our behavior may seem infinitely flexible, in reality it is not. In the end, we all do the same *sorts* of things, although individually, and as a society, we do those things in our own particular ways. More abstractly (as explored in recent years by Ken Binmore of University College London), all human individuals and societies, from the school playground to the chambers of the United Nations, have a highly developed sense of justice and fairness, and of individual dignity. Again, modern studies are beginning to show that other animals may also share such sensibilities: cats really do illustrate the importance of personal dignity, while many primates have a fine sense of social acceptability.

Each individual does things in his or her own way; each society has its own mores and traditions; and yet, once you get past the superficial differences, the similarities come roaring through. Again, this is a common theme of literature: people who thought they were quite alien to each other discover their shared humanity (like the Japanese soldiers, brutalized by too much war, and their prisoners in Nevil Shute's *A Town Like Alice*). Yet the question—Is there really something called human nature, or are we all just modeling clay?—has been a key theme of Western philosophy and politics for centuries. It has led to wars, and currently leads to unseemly spats in otherwise polite company.

The debate in its many forms is commonly summarized as "nature versus nurture." No harm would result if the nature-nurture debate was simply a matter for philosophers. Alas, however, the discussion has been extremely politicized. On the one hand, those who suggest that there is indeed something called human nature have often been accused of "biological determinism," or, if we assume for a moment that *biological* in this context means "genetic," of "genetic determinism." Enthusiasts for human nature are accused of proposing that each of us is merely a slave to his or her genome, slavishly behaving, minute by minute, according to the dictates of our genetic programs. The accusers suggest that this view precludes all ideas of free will or human conscience. As free will is cast aside, so too is personal responsibility. How can we be held responsible if we must simply do what our genes tell us, since we did not prescribe our own genes?

The notion of "genetic determinism" has perverted much of our history. It can be seen in the Old Testament in the story of Noah, when his various sons are decreed to be fundamentally different, with the descendants of Ham condemned to be slaves (Gen. 9:25). The church of South Africa justified the policy of apartheid by reference to Noah. Genetic determinism is thus the stuff of racism; it includes the notion that people of a different skin color (denoting different genes) are of a different clay and, ipso facto, inferior. It has also been the stuff of class distinction, which, as Marx said, has perhaps been the most divisive human proclivity of all. People born of poor parentage have, commonly, been assumed to be of no worth. Such notions underpinned the feudalism of Europe and Asia and have played a far greater part in the history of the United States than many enlightened Americans would be happy to admit, and still do. Of course, the notion that we are shaped ineluctably by our genes does not realize its full awfulness until it is combined with the notion that some genetic combinations are better than others: light skins better than black or brown, blond hair better than brunette, or worse, in some societies, "high birth" better than "low birth." But human beings have a great ability to attach value judgments whenever they detect differences. The willingness to say *"Vive la différence"* is a refinement of civilization advocated by the French, though not necessarily practiced by them. Genetic determinism is indeed awful.

On the other hand, those who have taken the tabula rasa position most literally have produced some of the most hideous and cruel societies ever perpetrated. As Christopher Badcock of the London School of Economics is wont to point out, the counterpart of biological or genetic determinism is "environmental determinism."

Stalin's USSR was based on the idea (crudely derived from Marx) that human beings would grow up happily as perfect communists, casting off their personal desires and ambitions in favor of the state, provided only that they were brought up in a perfect communist environment. Those who for recidivist and entirely reprehensible reasons did not find it quite so easy to conform to the state machine were seen to be anomalous—pathological—and were suitably disposed of.

In short, both the biologically determinist and the environmentally determinist positions are hideous, although at the time of writing the former does seem to be receiving the worse press. After all, we like to believe that we really can influence the way we are, and the fate of others, including our children, by manipulating the environment, so perhaps we are less alert to what can happen when the environment is manipulated to an extreme degree.

Fortunately, however, both extreme positions are nonsense. The sensible position is intermediate. Of course we have human nature, which in the end is rooted in our genes, but our genetic programs do not *prescribe* our particular actions, and they cannot. After all, modern Darwinism sees our genes as "survival machines": and they would be very bad survival machines indeed if they simply promoted inflexibility! We are shaped by our environment, but only within the—albeit broad—constraints of our innate biology. In truth, in all creatures there is constant dialogue between the genome and the environment: each proposing; each disposing. In human beings the dialogue is particularly complicated because our genes, over evolutionary time, have produced great brains, which give us a degree of individual flexibility that far exceeds that of other creatures. But each of us is nevertheless a dialogue between genome and environment.

Since Darwin's (and Mendel's) day there have been several distinct attempts to discuss human nature—or at least human behavior—in evolutionary and genetic terms. These attempts have given rise to a series of quite different schools, with different premises, approaches, and ambitions. Critics of evolutionary psychology—or what they seem to imagine is evolutionary psychology—tend to conflate the different schools. Before we begin, it seems worthwhile to outline the distinctions.

First, there is a school that derives from Francis Galton's biometrics. In general, it has sought to explain *differences* between human beings in genetic terms. This school, for example, has given rise to the various exercises to rank people's intelligence, measured as IQ. Evolutionary psychology is not concerned with this. It is concerned with universals: the things that human beings have in common; the qualities that link us all. Thus an evolutionary psychologist would study the human ability to acquire language—contrasting this extraordinary skill with the language abilities of other species—but would have no interest in collecting data to see whether some people are more linguistically adept than others. However, by seeking out the biological roots of language, evolutionary psychologists have managed to show that some specific inabilities, like dyslexia, have specific causes. Teachers and parents of dyslexic children—and the children themselves—are not distressed by this. On the contrary, the revelation of specific dysfunction has been liberating—removing the blame and the charges of idleness that have often dogged sufferers in the past, removing the guilt from the parents, generally putting everyone in a more positive frame of mind, and also (which in the end is the point) suggesting new approaches to treatment.

The differences between human beings that evolutionary psychologists *are* concerned with are those between men and women, or boys and girls, or children and adults. It is cruel and destructive to assume that the rights and desires of one particular category of people in a society should prevail: typically, the rights of aging males. But it is also cruel and destructive to begin with the premise that all human beings are the same, and to assume that everyone in a society really does have the aspirations and propensities of an aging male, or would have, if only they were given the chance. It seems far more sensible and humane to recognize differences where they exist and then build societies that enable all aspirations and abilities to flourish.

Then, in the mid–nineteenth century, Herbert Spencer proposed the philosophy known as social Darwinism, though in fact he began developing it before Darwin published *Origin of Species*. He attempted to explain the structure of society in terms of natural selection. But he made two mistakes. First, his view of natural selection was crude and Victorian, nothing like the modern view. It saw nature as one long battle, reflecting Tennyson's pre-Darwinian "Nature, red in tooth and claw." Spencer, not Darwin, coined the expression "survival of the fittest." Second, he supposed that since nature itself leads to the survival of the fittest, then that is how human societies *ought* to behave. Thus he slides seamlessly from a description of how life *is* (not that it really is, but he thought it was) to a recommendation of how life *ought to be*. David Hume pointed out that philosophers often do this—slip without pause from "is" to "ought"—and that this is a very sloppy thing to do. There is no necessary connection, he said, between "is" and "ought." In *Principia Ethica* in 1903, the English philosopher G. E. Moore took the point further. He coined the expression "naturalistic fallacy," by which he meant that it simply is not permissible to argue that what is natural is necessarily right. Ethical principles should be framed *without* reference to what nature happens to do. Lions practice infanticide on a massive scale, but that does not make it right. So Spencer's social Darwinism was wrong on two counts: its view of nature was misguided, and he fell foul of the naturalistic fallacy. It is amazing, however, how many modern critics of evolutionary biology apparently conflate it with Spencer's social Darwinism.

However, Moore's naturalistic fallacy need not be taken as the last word. In truth, we *do* take nature into account when framing ethical principles, and it is surely not unreasonable to do so. In the debates on human cloning and in vitro fertilization, for example, we tend to begin with the premise that it is acceptable for young women to want babies because it is "natural." If we did not feel that childbirth is good because it is "natural," we would not debate those technologies at all.

Sociobiology is the late-twentieth-century attempt to apply evolutionary ideas—*modern* evolutionary ideas, not those of Tennyson—to human behavior. E. O. Wilson suggested the term in the early 1970s. Evolutionary psychology is the direct descendant of sociobiology and in most respects is indistinguishable from it. The change of name was largely for political reasons. Some early sociobiology enthusiasts simply forgot—or were never aware of—Moore's naturalistic fallacy. They pointed out that men are able to produce more offspring than women and then went on to argue—or apparently to argue—that *therefore* it was permissible for men to philander because it was "in their nature" to find as many mates as possible. By contrast, women could not physically produce more than a few children in the course of their lives and needed above all to find men to help them bring up the children, and therefore should be monogamous. As in Spencer's case, much of the underlying biology is suspect, because the discrepancy in fecundity between men and women is nothing like so great as might at first be expected. No man is known to have had more than about 850 offspring, while the maximum known offspring for a women is over 60. Much more important, however, the naturalistic fallacy must apply. Even if it is in a man's biological interest to impregnate as many women as possible, this does not make it right.

Because of this philosophical lapse, sociobiology in its initial form got itself a bad name. The renaming—as "evolutionary psychology"—was an attempt to continue the good work that was implicit in sociobiology but to shake off the bad philosophical impedimenta. Modern evolutionary psychologists commonly begin public lectures by acknowledging the naturalistic fallacy: "The way I suggest that human beings *are,* is not the way I necessarily think they ought to be."

Evolutionary psychology will, I believe, prove to be one of the greatest intellectual legacies of the twentieth century. It is unfortunate, however, that some critics of the subject are unwilling to engage in serious argument and that old accusations like "genetic determinism" continue to be published. In the twenty-first century and beyond, evolutionary psychology will surely prove to have immense implications for all human affairs: medical practice, sociology, moral philosophy, and politics. In this light, then, we must ask what the subject is really about.

THE ESSENCE OF EVOLUTIONARY PSYCHOLOGY

All ideas in science prompt us—or should prompt us—to ask three questions: Is it plausible? Is it testable? And if it were true, what difference would it make? We will look at these issues as they arise, but the general answers regarding evolutionary psychology are, I suggest, "Yes," "Yes," and "In many contexts, a great deal."

At present, many different and often apparently unrelated notions are being discussed under the general heading of evolutionary psychology because it is still a young subject, and all subjects take time to settle down. But the idea as a whole is rooted in four fundamental premises that are worth looking at one at a time.

Premise Number One: *Genes do, to some extent, underpin the way we behave.*

Some people, including members of the scientific community, find even this most basic proposition implausible. They argue that genes are merely strands of DNA, while behavior is, well, behavior—and how could one possibly lead to the other?

People in general and scientists in particular tend not to take ideas seriously unless they find them plausible. The crows in Walt

Disney's *Dumbo* were quite right to dismiss the rumor that an elephant might fly. It is just not plausible. Their ears could never be big enough. They could never generate the power. Cartoons are weightless, but real elephants must respond to the call of gravity.

Yet we should not set *too* much store by plausibility. If we did, we would still be in the Middle Ages. Much of the job of science has been to show that unlikely ideas are indeed true: that the universe is not always as it seems. Darwin made the idea of evolution respectable largely because he provided a mechanism—natural selection—that seemed to make it plausible. With natural selection, it was possible to see how intelligent creatures could evolve from less intelligent creatures, and less intelligent creatures might evolve from entities that had no intelligence at all. Many biologists before Darwin had proposed that evolution was a fact, but none had provided a plausible mechanism. After Darwin, most biologists believe that evolution is indeed true, and that natural selection is the most powerful shaping force within it. But is it the plausibility of natural selection that makes evolution true? Of course not. Animals and plants were happily evolving for hundreds of millions of years before Darwin, even though no one had yet conceived of a plausible mechanism.

Similarly, in 1915, Alfred Wegener proposed continental drift: the notion that the continents have shifted around the surface of the globe, such that South America, for example, was once joined to Africa. At first geologists were intrigued, but then they laughed the idea out of court. They simply could not see how this could happen. Continents could not plow their way through the oceanic crust as a ship can plow through the ocean's waters. Ergo, they said, continental drift is a silly idea.

Then, in the 1960s, geophysicists formulated the notion of plate tectonics: showing that the Earth's crust itself is divided into sections, and that these move relative to each other, carrying the continents with them. This made continental drift plausible, and of course, it is now accepted as the orthodox, scientific truth. But is it the plausibility that makes the idea true? Not at all. Continents were indeed drifting in 1915, even though Wegener himself did not suggest a plausible mechanism—they have been shuffling across the surface of the globe for more than 4 billion years, when the Earth was first consolidated.

In short, we are right to demand plausibility. The criterion of plausibility saves us from an uncritical pursuit of flying elephants. But we have to be prepared to be surprised. All the best ideas in science—including that of the gene—seemed implausible at some point, to somebody. So when a professor of biology tries to tell us as a matter of dogma that genes *cannot* underpin behavior because, well, genes are merely chemistry while behavior is behavior, we might pause to spare him the time of day but then remind him that plausibility, in the end, is a poor criterion of truth.

But if we look more closely, we realize that, implausible or not, *of course* genes influence behavior. Chimpanzees do not behave like dogs, or dogs like blackbirds. Of course they are anatomically different, so they cannot all behave identically. Yet anatomy is part of the point: no one doubts the genetic basis of anatomy, or that it influences behavior. But then again, animals that are very similar anatomically may still behave quite differently. Some birds are monogamous (or nearly so), and some appear so but are highly unfaithful, while others of similar constitution may be polygamous and others are communal breeders. They are all anatomically capable of adopting each other's modus vivendi, but they do not do so. Have their genes really got nothing to do with this?

Of course, some people just like to think that human beings are not subject to the normal rules of biology. In essence, this is an ancient idea: it is the anthropocentric Judeo-Christian notion that we alone, of all creatures, are made in God's image and so are qualitatively distinct from all the rest, which are merely his creatures. Nowadays, orthodox Judeo-Christianity is less fashionable than it once was, but the idea persists in new guises, one of which is the notion that any suggestion of genetic influence is a species of "genetic determinism," which is ipso facto bad. This is the view that is now politically correct. Biologists who choose for whatever reason to go down the politically correct route point out that human beings have enormous brains, which, they suggest, must override the cruder input of the genes.

Are they suggesting that those brains were made *without* input from the genes? Presumably not. Presumably, as biologists, they must accept that genes make brains in the same sense as they make stomachs and pancreases and that the function of a brain (just like the function of a stomach or a pancreas) is influenced by the way it is put together. This is all that an evolutionary psychologist would claim, or needs to claim. Genes make brains. Nobody claims that the DNA itself tells the person who contains it what to do. That really would be implausible, not to say silly. Nobody claims that the layout of each and every synapse in the brain is prescribed by some particular, preassigned gene. But then nobody claims that the release of every single red blood cell in the bone marrow is prescribed by its own particular gene. Genes do not work like that. No one ever said they did. Genes work at longer range. They set processes in train by making proteins that initiate chains of metabolism. In the case of animals (including humans), evolutionary psychologists merely claim that genes make brains that can think and initiate actions. That is not implausible. That is simply what happens. Deny that, and you are denying the essence of all biology.

Finally, who would reject the notion outlined at the head of this chapter: that a human brought up by giraffes would never feel quite at home? That he or she would never quite master the behavioral repertoire of the giraffe, and yet would feel constrained by it? Would anyone really want to deny that the sense of discomfort would derive, at least in part, from the genetic difference between human beings and giraffes?

In short, the first premise should not be controversial. Of course our genes influence the way we behave. The general question is,

To what extent do they do so? More particularly (although evolutionary psychology is more concerned with human universals than with human differences), To what extent can behavioral differences between, say, men and women, or children and adults, be ascribed to genetic differences? Then again, How do genes influence our behavior: What are the mechanisms? And, To what extent, by the exercise of what we like to call free will, can we override the influence of our genes? These are all legitimate questions. But the premise on which they are all based—that our genes are in there somewhere—is, when you think about it, simply undeniable.

Premise Number Two: *The selfish gene.*

I said at the end of the last chapter that the selfish gene—an expression coined by Richard Dawkins, although he never laid claim to all the ideas behind it—is the most significant concept to emerge in evolutionary theory since the modern synthesis was put together. It is relevant in all evolutionary contexts, but we can perhaps see its power most vividly in the context of behavior.

Natural selection is the mechanism of evolution, but at what *level* does it operate? Darwin had no doubt that natural selection picks out the *individuals* that are best adapted to the prevailing circumstances. The "fittest" individuals are the ones that are most likely to survive and to produce offspring that resemble themselves.

Others later (as Darwin himself had done) observed a great deal of behavior in creatures of all kinds that did *not* seem necessarily to favor the individual who was doing the behaving. Often creatures seemed to behave unselfishly. The most obvious example is the way that parents of all species—plants as well as animals—so often sacrifice themselves on behalf of their offspring. This example is so obvious that most biologists overlooked the innate oddness of it, that an individual can sacrifice himself or herself for some other individuals. There were many examples, too, of apparent sacrifice for individuals other than one's own off-spring. If natural selection favored the individuals who strove most effectively to survive and reproduce, how could it ever favor unselfishness?

The problem seemed to have been solved in the early 1960s by a British zoologist, Vero Wynne-Edwards. His book of 1962, *Animal Dispersion in Relation to Social Behaviour,* described most famously the behavior of red grouse. The males compete fiercely for territory, but if they fail to obtain a territory after a reasonable time, they simply give up, sit on the margins of other males' terri-tories, and (usually) quietly pass away. Why? On the face of things, they have nothing to lose by continuing the fight. If they go on challenging for more territories, they *might* acquire one and go on to raise families of their own, but if they simply give up, they are *bound* to die without issue. It seems, therefore, that natural selection ought to favor red grouse males that are prepared to fight to the death. But instead, natural selection seems to have favored the quitters. It's all very puzzling.

Wynne-Edwards proposed the notion of "group selection." Natural selection, he proposed, does not in fact operate at the level of individuals, as Darwin had thought. It works primarily at higher levels—in effect, at the level of the breeding population as a whole. Male grouse that fought all out, beak and claw, for terri-tory might indeed succeed sometimes, and might be favored over males that simply gave up. But if all males always fought to the death, then the breeding population as a whole would be wrecked. Every male that had a territory would have go on fighting and fighting and fighting until the last dispossessed individual had been slaughtered. By that time the breeding season would be over, even if anyone had any energy left to look after the babies. No: it would surely be much better all around if animals tempered their behavior in favor of the group as a whole, assessing how many could breed, fighting for their own interests as long as there was a reasonable chance of winning, and giving up when the game was obviously up. Natural selection, then, would surely favor creatures that were alert to such subtleties.

But this idea—grand, ennobling, ingenious, and, for a time, influential as it was—finally fell at the hurdle of plausibility. It was easy to see how such a group selection mechanism would have advantages, how once it was in place it really could help the population to survive better than it would if nature really was as red in tooth and claw as Tennyson had proposed. The idea did indeed seem to fit the facts. But it was impossible to see how such a mechanism could evolve in the first place. If all the individuals in the group played the game in the way that Wynne-Edwards proposed, the system would be efficient and would surely persist. But what if some mutant grouse was born who really did prefer to fight to the death? Surrounded as he would be by such gentlemanly grouse, he would have a relatively easy time. If he went on battling while others gave up, he would surely acquire territory and produce offspring like himself, battling bruisers. Soon the population of gentlemen would be overrun by hooligans. To be sure, the hooligans would probably die out. To prevent this happening, however, natural selection would have to be able to look ahead; it would have to include mechanisms that said, "Don't behave in an ungentlemanly fashion, for if you do you will eventually endanger the group as a whole, including your own children." But a grouse that thought in this way would lose out to one that said, "To hell with that! If I fight my way in now, I can have children of my own, and if I just lie down and die, I won't!" Thus the ungentlemanly behavior would win in the short term, and unless you win in the short term, there is no long term.

So now biologists were back where they started. If natural selection operated primarily at the level of the individual, as Darwin argued, then it was extremely difficult to explain the many cases in which animals (or other organisms) seem to behave self-sacrificially or self-effacingly, as with the male red grouse. Wynne-Edwards's idea—that natural selection operated on the group as a whole—did not seem to work. Self-sacrificial behavior, loosely classed as "altruism," seemed impossible to explain. But thousands of observations proclaim that it is common. Once we perceive that parental investment is an example of altruism, with or without additional childcare, then it becomes universal.

The answer was worked out in the late 1960s and 1970s primarily by Bob Trivers and George Williams in the United States and by William Hamilton in England. Again, we must envisage a shift in the level at which natural selection operates most forcefully. Paradoxically, though, we should not shift the focus of our attention upward, from individual up to group. Rather we must shift downward, from individual down to gene. The message that came roaring from the 1960s and 1970s, in short, was that *natural selection operates primarily at the level of the gene.* Individuals can be seen merely as vehicles for the individual genes that they contain. This is the idea that Richard Dawkins summarized and developed so cogently in *The Selfish Gene,* in 1976. The shift of focus from individual to gene is the most significant intellectual advance in evolutionary theory since the formation of the modern synthesis itself.

The paradox—or the apparent paradox—is that if genes are "selfish," then it becomes easy to see how the individuals that contain them can behave unselfishly, or, as biologists say, "altruistically." Suppose there are two female rabbits. One possesses a gene that tells her, "If your babies are attacked, fight with might and main to defend them!" The other possesses no such gene. When the polecat comes to the first rabbit's burrow, she fights like mad and drives him away. Her eight babies all survive, and each of them, in turn, has inherited the gene that says, "If your babies are attacked, defend them!" When the polecat comes to the other rabbit's burrow, she sneaks away, and all her babies perish. Of course, sometimes the first rabbit will be killed and so will her babies, but in general, we can see how defense of young beyond the normal course of duty must enhance their survival, and that risk of death is a reasonable price to pay for this. We can also see, which is much more to the point, that a gene that told its possessor to fight for the survival of the offspring would spread through the population. Such a gene would, after all, be inherited by the survivors. A gene that simply favored cowardice would die out as the children were slaughtered.

Of course, the above case is simplistic. Genes do feed into behavior, but no one gene operates alone, and all in practice underpin mechanisms that have conditional clauses built into them. So a gene that encouraged its owner to fight for her children's lives would not normally be expected to say, "Sacrifice yourself at the drop of a hat on every occasion!" We see in nature how mothers of all species effectively weigh the cost and the risk of fighting against the option of conserving their own lives and producing more babies later. This appalling dilemma is evident in the body language of any lapwing as she attempts to lure predators away from her chicks, at great personal risk but usually falling short of sacrifice. On the other hand, birds in general are long-lived creatures that *do* reproduce repeatedly. Other creatures, liable to reproduce only a few times, might do more to defend any one litter. These are refinements, however. The general points remain. First, a gene that tells its possessor to defend her young will spread through the population. That is what is meant by being favored by natural selection. Second, once we see that natural selection operates on genes more powerfully than on individuals as a whole, we see how it can indeed favor unselfish—altruistic—behavior. Thus the problem that Darwin felt so acutely is solved at a stroke.

Of course, the sacrifice of parents for their offspring is a special case, albeit an almost universal special case. In practice, biologists have identified three main sets of circumstances in which altruistic behavior would be favored by natural selection.

The first is *kin selection*. The care of parents for offspring is the principal but not the only example of this. The basic point is that a gene will promote its own replication if it prompts the creature that contains it to lay down its life to save other creatures that contain the same gene. A few conditions have to be fulfilled for this to work. First, the number of individuals saved has to exceed the number sacrificed. Second, the creature that is sacrificing itself has to be pretty sure that the creature for which it is making the sacrifice really does contain the gene that is promoting the behavior. There is a 50 percent chance that any one gene contained in the parent is also contained in any one offspring; and a 50 percent chance that a gene contained in any one sibling is contained in another sibling. There is a one in eight chance that a gene contained in any one individual will be contained in a cousin. In general, then, we would expect individuals to sacrifice themselves for their offspring or siblings—and also to sacrifice themselves for cousins, but to make less of an effort. In general, we would expect individuals to behave in ways that promote the welfare of their own families. Bill Hamilton extrapolated from this the idea of *inclusive fitness.* Natural selection will favor a gene that promotes acts that maximize replication of that gene. Replicates of that gene are found in the individual's relatives: offspring, siblings, cousins, aunts. The sum total of benefit to all the copies of the gene in all the relatives is called inclusive fitness.

Kin altruism, leading to inclusive fitness, is a kind of one-step operation. By promoting a particular kind of behavior in the individual that contains it, the gene in effect replicates itself. Other mechanisms that promote altruistic behavior require some measure of reciprocity. An individual is nice to another individual (or group of individuals) who in some measure returns the favor. The gene that promotes the altruistic behavior does not automatically become replicated by doing so (as is the case with kin altruism). But it does help create conditions that enhance its own chances of survival and replication.

The two well-recognized examples of this are mate selection and reciprocal altruism. Animals might well be nice to other animals that they hope to mate with, or indeed have already mated with. The mating partner, after all, provides the means for the replication of their own genes. The apparent unselfishness is well worth it. "Reciprocal altruism" implies exactly what it sounds like: a creature does a favor for another creature in the hope and expectation that, on some future occasion, the favor will be returned. There are several prime examples in nature. One involves vampire bats. They feed by lapping blood from other animals, such as donkeys and cattle, which is a somewhat precarious way to live. Sometimes the bats get too much to eat, and sometimes too little. On any one night, those that have too much will regurgitate some of the surplus for their less fortunate nest-mates who have got too little. The expectation is that on some future night when fortunes are reversed, the favor will be returned. Another example involves the cooperation of subdominant males in some apes and monkeys. One of the cooperators engages the attention of the alpha male while the other one mates with one of the females. Again it is expected that on future occasions, the favor will be returned. Finally, gray whales practice cooperative matings. Copulation in open water is not too easy, and one male will help another to a successful conclusion. He expects the favor to be returned, however.

As Matt Ridley points out, however, convincing examples of reciprocal altruism in nature are not common. He prefers a more general suggestion, which at least applies to human beings: that natural selection ought to favor trading. There are not many circumstances in which any one creature is liable to be able to repay a favor *exactly,* within an interval of time short enough to remember. But if the animal that had once been favored were simply to return some equivalent—some quid pro quo—then this would be trading of a kind.

My own contribution to world wisdom is to suggest that natural selection ought generally to favor genes for conviviality. The logic is as follows. Natural selection clearly favors social life, at least in many creatures. If it were not so, there would be no social animals. Given that many animals are social, they need to get along with their fellows, and indeed, in social animals a huge proportion of the behavioral repertoire (and, very often of the brain itself) is occupied with social behavior. It seems to me self-evident that natural selection should favor social creatures that invest at least some of their time in being nice to other members of their society, whether or not they expect any immediate favors. In the end, this is simply a way of lining your own nest without having to try very hard. Life is certainly a lot easier if you are surrounded by friendly rather than unfriendly people. Friendliness breeds friendliness, and enmity breeds enmity. In short, a gene that favored general friendliness (unless the creature was otherwise provoked) should surely be favored.

Premise Number Three: *Behavior has indeed evolved by natural selection, and many aspects of our behavior were adaptive once, even if that no longer seems to be the case.*

This is where the term *evolutionary* comes in. We begin with the first premise (that genes may underpin behavior) and then ask why a particular creature should find itself with the genes that it has in fact got. As good post-Darwinians, we assume that at least a significant proportion of a creature's genes have been selected through the course of evolutionary history because they conferred some survival or reproductive advantage. A horse has legs that equip it for long-distance travel because a life as a grazer on the open plains would be aided by such equipment; a peregrine is sharp in beak and claw because this favors life as a predator. The assumption in the present context is that we have genes that shape our behavior in particular ways because at *some time in our evolutionary past,* natural selection favored such genes. In accord with premise 2, we would assume in general that natural selection would favor genes that promoted behavior that favored their (the genes') own replication.

Two interesting and perhaps somewhat contrasting points follow from this. The first is that our behavior—or moods and thoughts—may well have some adaptive advantage even when this is not obvious to us. But the second is that particular sets of genes have in general been selected in conditions that prevailed in the past but no longer prevail. In other words, some at least of the genes that shape our behavior may have been adaptive once but now no longer confer any advantage, or indeed may now be maladaptive.

Simple examples of "behavioral genes" that were favored by conditions in the past but are no longer pertinent include some that influence our dietary preferences. Thus, the hypothesis has it, in the Pliocene and Pleistocene epochs (between 5 million and, say, 40,000 years ago) our ancestors lived in the savanna and woods of Africa. There it would have been difficult for them to obtain enough energy for day-to-day life, or indeed enough sodium, normally eaten as sodium chloride (common salt). Natural selection therefore favored individuals that had a particular liking for high-energy foods and for salt, and would therefore go to considerable lengths to seek them out. In the wild, high-energy foods include honey and fat, and not much else. Both are relatively rare (wild game is not generally very fatty) and well guarded. Bees guard honey, and wild animals do not give themselves up easily. But in a modern agricultural-industrial society we find it very easy to produce fat livestock, we produce sugar by the megaton, and we mine salt. In short, modern people are surrounded by plenty, but they have inherited the appetites of ancestors who lived for many generations in extreme austerity. Here then is an evolutionary explanation for the common observation that human beings, given the chance, so readily eat to excess, even though, as a result, they suffer all the "diseases of affluence," from obesity to heart disease and diabetes.

There are parallels among "behavioral genes," too, for the many cases we noted in earlier chapters: of genes that are useful in some genetic contexts, and less useful in others. The classic case is the sickle-cell gene, which protects against malaria in a single dose, but leads to damaging anemia in a double dose. It now seems clear that autism has a genetic basis. It also seems to be the case, however, that the gene "for" autism also enhances its possessor's spatial sense.

The notion that the genes that shape our behavior have been selected through our past experiences is at the core of evolutionary psychology. Clearly it has given the discipline its name. However, this key idea offers leverage for criticism. Once you get into the rhythm of the idea, it can be very easy to think of "evolutionary" explanations for just about any kind of behavior—or any other quality, behavioral or not!—that we possess. Stephen Jay Gould, who criticizes evolutionary psychology on various grounds, has used the expression "just-so stories" to describe all the ad hoc, apparently "evolutionary" ideas that people dream up to "explain" just about anything.

However, serious evolutionary psychologists are serious thinkers. They are fully aware that science is not science until it produces specific hypotheses that are testable. Its practitioners do not merely think up just-so stories. They frame their speculations as testable hypotheses, and they undertake the specific tests. But there are two snags. First, evolutionary psychology is not yet fashionable, so it does not attract an enormous amount of grant money, and research cannot be undertaken these days without grant money. Second, there are so many variables to be tested when trying to explore human behavior that it can take many years to produce robust results. Thus Margot Wilson and Martin Daly have taken more than 20 years to demonstrate beyond any reasonable doubt that stepparents are at least 100 times more likely to kill their stepchildren than "natural" parents are to kill their own children and that the critical problem is not one of poverty, or social class, or general social instability, or any one of a dozen other possibilities that have often been suggested, but is primarily one of genetic relationship. Since the whole discipline of evolutionary psychology is new, and good studies can take such a long time, it is hardly surprising that, so far, so few have been done.

As long as ideas remain untested you can, if you really want to be derogatory, call them just-so stories. However, there is a significant difference between a serious idea that could be tested—given time and resources—and one that is merely intended to amuse (like Kipling's original *Just-So Stories*), and is perhaps not testable at all. It is mischievous to place serious ideas that haven't been tested in the same category as ideas that may not be testable, and even if they were, were only intended to be frivolous in the first place.

The fact is, though, that the most crucial thread of the whole thesis—the idea that our behavior has been shaped by natural selection, and was adaptive once even if that is no longer the case—is in principle the most difficult to demonstrate. That is simply a cross that has to be borne and, eventually, overcome.

Premise Number Four: Quantification.

Strictly speaking, of course, "quantification" is not a premise: it is merely a modus operandi. But it is fundamental. Evolutionary theory has grown on the back of a series of mathematical advances that make it possible to measure, compare, and assess different behavioral strategies. The general notion is that natural selection will favor the strategies that are most likely to enhance survival and reproduction (meaning the survival and replication of the gene that promotes the behavior).

Mathematicians use a number of devices to assess the value of different behaviors, but two particular kinds are outstanding. First, there are various optimization models, whereby the scientist first seeks to work out what the creature is actually trying to do, then works out mathematically how it might best solve the problem, and then tries to see how closely the animal's behavior conforms to what the math says is ideal. The scientist does not assume that if the animal is not carrying out what the math says is ideal, then the animal has got it wrong. Probably—much more to the point—the animal knows more than the analyst does and in fact is trying to do more than one thing at once. Optimum foraging theory is a good example of the whole genre: asking what an animal needs to do to maximize its food intake at a given time. Blackbirds feeding from cotoneaster berries seem to hop from branch to branch, or from bunch to bunch, more frequently than they strictly need to, and this may seem a gratuitous waste of time and energy. However, the blackbird also needs to avoid its own predators, including cats and hawks. So it stays on the move as it feeds. Optimum foraging theory, then, properly applied, not only helps us see how efficiently an animal is functioning, but also helps us perceive more clearly than we otherwise might the nature of its problems. We can reasonably assume that all creatures—including humans—operate in large part according to optimizing strategies of one kind or another. It is hard to see how we would still be around if that were not the case.

The second kind of model that is proving particularly instructive is game theory. This pits a creature not simply against its environment, as optimization theory does, but against others of its kind. One example of game theory in action is the prisoner's dilemma, in which two players gain maximally if they betray each other and lose maximally if they are betrayed, but each does reasonably well if only they both cooperate. The question is, if you don't know how the other player is going to behave, should you begin by cooperating or by betraying? If you betray and the other person plays fairly, then you win everything and the other loses everything. If you cooperate and the other player cheats, then you lose everything. On balance, it seems safest to cooperate—but of course, not necessarily.

The prisoner's dilemma in a sense encapsulates much social life. Should we set out to cooperate with our fellows and risk losing out completely? Or should we "get our retaliation in first," as villains like to say, and operate by a policy of hit and run? In reality, of course, few confrontations with our fellow human beings are once-and-for-all, never to be repeated. In normal social life we meet the same people over and over again. So we acquire reputations. If we acquire a reputation as a cheat, nobody will cooperate with us, so we can never pull a fast one on anybody else. If we acquire a reputation as a soft touch—as someone who never retaliates, no matter what the provocation—then we will always be taken advantage of. In general it seems best to be known as someone who is fundamentally honest and cooperative but is not to be trifled with. All this emerges from the simple application of game theory. It is remarkable how closely it conforms to the way human beings behave in practice, and how carefully they build and guard their reputations.

These four kinds of notion, then, lie at the root of evolutionary psychology: that genes do indeed influence behavior; that natural selection operates primarily at the level of the gene; that our fundamental behavior patterns have evolved through natural selection and are likely to have conferred an adaptive advantage at some time in the past, even if they do not seem to do so in the present; and that various mathematical models can be used to predict likely behavioral strategies. However, the notion that our behavior does have deep and analyzable roots in no way denies what is obvious: that we also have brains that enable us to remember, to think, and to communicate; and that whatever we do has a heavy overlay of culture. What remains fascinating is the extent to which apparently huge cultural differences tend to emerge as variations on underlying, universal themes: language, childcare, courtship strategies and mate choice, and so on. To understand human beings, surely, we need to see both the cultural differences and the underlying biological similarities, and to see how each plays upon the other. Anthropologists and psychologists in the twentieth century have focused largely on the cultural differences. The attempt by evolutionary psychologists to describe the underlying similarities can be seen as an attempt to restore the balance, no more, no less. It is not, innately, a threatening pursuit, nor is it a fatuous one.

These, then, are the basic premises and modus operandi of evolutionary psychology. A number of key themes have so far emerged from the subject. We can place these themes in three general categories: strategies for living, how the brain works, and specific quirks and pathologies.

STRATEGIES FOR LIVING, MATING, AND REPRODUCING

The essence of life is to survive and reproduce, and to both these ends most creatures have evolved some measure of sociality. Evolutionary psychology is concerned with all three components of life: survival, reproduction, and sociality.

Most creatures that reproduce sexually evolve two distinct sexes. (They may evolve more than two different "mating types," but on theoretical grounds alone, two is the most likely number.) Males produce a great many very small gametes that are highly mobile (spermatozoa), and females produce a much smaller number of big gametes, laden with nutritious yolk, that are immobile (eggs).

This difference in physiological output immediately suggests, and prompts, a difference in reproductive strategy. Each spermatozoan is capable of giving rise to a new individual, so that each male even of unfecund animals, such as cattle and humans, could (so crude arithmetic suggests) produce hundreds or even many thousands of offspring.

Females, on the other hand, must invest a great deal more in each individual egg than a male does in each individual sperm. An egg, typically, is many thousands of times larger in weight and volume than a sperm. Simply because she invests so much in each individual egg, the female *must* invest far more *per individual offspring* than the male is bound to do, and as a result of this she cannot produce anything like so many per lifetime as the male can.

Given these raw, physical facts, we would *expect* males and females to have different behavioral and emotional approaches to reproduction. Broadly speaking, we would *expect* males to focus on quantity. Crude arithmetic alone suggests that a male that produces a thousand offspring spreads its genes more effectively than one that produces only two, or, more to the point, that a gene that prompts its owner to be as fecund as possible is more likely to spread through the population than one that encourages its possessor to exercise restraint.

On the other hand, we would *expect* females to focus on quality. A female human being living as a hunter-gatherer probably produces only five babies in the course of a lifetime. Other very large animals—rhinoceroses, elephants, orangutans—probably manage about the same number. Some of the babies will die at birth, some will suffer accidents, but with luck, two out of the five will survive to have offspring of their own. Obviously, if each female produces two successful offspring that survive to reproduce, then the population is sustained. This is not *why* females tend to produce two successful babies per lifetime. But if they do not, then their lineage goes extinct. Since she has only five or so babies in the course of a lifetime, we can expect the mother to lavish enormous care on each individual, far more than a male might consider worthwhile.

Common sense suggests that the innate difference in reproductive strategy should operate at two levels. First, we would expect different mating strategies. Broadly speaking, we would expect males to seek to impregnate as many females as possible, and so to pay less attention to the perceived quality of each one. But we would expect females to be as choosy as possible. In the course of a lifetime they will entrust their genes to only five partners at most, or five, that is, who will actually furnish them with offspring. So they want the best possible partners. Furthermore, we would expect females to be looking for qualities in their partners that would be likely to bring the greatest reproductive success in their children. The question is often asked, in life and in literature, "Why are women so often attracted to 'rotters'?"—where *rotters* are defined as men who treat women badly, on a "love 'em and leave 'em" basis. The answer is that the son of a rotter inherits the rotter's genes, and is comparably promiscuous. So a gene that encourages promiscuity in a man is (by and large) spread more widely; if a woman hitches her own genes to the genes of such a man, then her genes will be widely spread as well. Genes are indeed selfish, and do not care about the happiness, or other such indulgences, of their possessors.

Second, we would expect mothers to invest more in their babies after they are born than a father commonly does. Of course there are many extremely indulgent fathers, in birds as well as among humans. But we comment approvingly when males are indulgent fathers. We *expect* mothers to be devoted. When they are not, we feel this is "unnatural."

Of course, in musing along these lines, it is easy to fall into deep philosophical traps. Some advocates in the history of evolutionary psychology have done so. But the critics commonly accuse the advocates of falling into the traps even when the advocates demonstrate conscientiously that they have avoided them, such is the zeal of the critics. The first trap is that of oversimplification. To critics of this chapter let me beg the favor of observing that I am myself drawing attention to this problem, and kindly note that this is an extremely abbreviated account. So although I point out how "crude arithmetic" suggests that a male should be promiscuous, I do not suggest that this is, in practice, the best thing for a male to do. In fact, among humans it is really very surprising how *few* offspring men have in the course of a lifetime, even when absolutely everything is lined up in their favor. Middle Eastern potentates sometimes ran reproductive industries, with fertile young women selected and primed nightly for their attentions.

Surely they could easily produce thousands of offspring in a life-time's endeavor, several hundred per year for 30 years or so. Yet the highest recorded number is around 800. In short, common observation shows us that "crude arithmetic" does not apply crudely in nature.

Of course, we know too that among human beings—the species that interests us most—it generally pays a father to look after a few offspring well, rather than produce illegitimate children in the far-flung corners of Earth who may come to bad ends for want of parenting. So we should not be surprised if in fact human fathers in modern societies behave toward their offspring much as the mothers do: indulging each one, and *not* spreading their seed indiscriminately. More broadly, we find that rape—forced promiscuity—is not a particularly common reproductive strategy in nature even though crude arithmetic might suggest that it would be. But monogamy—including faithful mating for life—is relatively common, especially among birds. In reality, then, a great many complicating variables override the crude arithmetic even before we add any ethical considerations.

The second trap is the one we outlined above: that of the "naturalistic fallacy." We must indeed acknowledge the arithmetic that underpins evolutionary theory, but it is crude in the extreme to allow the arithmetic to shape our morality. All serious evolutionary psychologists acknowledge this. To deny that they do so is simply misrepresentation.

In truth, we seem to face a dilemma. On the one hand, many a society simply applies stereotypes (which itself is bad) and then for good measure falls foul of the naturalistic fallacy. That is, to put the matter in simplistic terms, men are perceived as nature's feckless idlers, boozing, philandering, and gambling; while women are perceived as natural Marthas, faithful, house-proud, washing diapers and waxing the kitchen floor. In contrast, it has become politically correct simply to maintain that there *are* no innate differences between men and women, that all perceived differences are "contingent" (the fashionable word) on circumstance. If only girls were treated like boys, they would grow up more boylike, and vice versa, and that is assumed to be a good thing. Of course upbringing has an influence. On the other hand, it is undeniable that boys left together tend to wrestle and compete, while girls more generally converse and cooperate. A billion children's bedrooms and a million school playgrounds worldwide really can't be wrong. To be sure, it is cruel to assume that boys should all behave like Wild Bill Hickok, and girls like Little Dorrit. But it is at least as cruel, and certainly procrustean, to assume that both should necessarily behave like some synthesis of the two. A good and humane society, surely, is one that recognizes differences—including those that are innate—and reflects and accommodates the abilities and predilections of all.

Helena Cronin, a biologist at the London School of Economics and the author of *The Ant and the Peacock,* has suggested how the dilemma can be resolved. The point, she says, is simply to expunge all traces of the naturalistic fallacy from the discussions and, more broadly, all the accompanying value judgments. For example, many studies suggest that women bring different values to the workplace than men. They do not set such store, as men do, by status and income. This is not a criticism, either of men or women, but it is a fact, insofar as facts about people's attitudes can be assessed by asking them what they think. If anyone (man or woman) sets less store by status and income, then he or she will compete less vigorously for high-status, high-income jobs. In some sectors, the high-income jobs are also the most risky, which is why they are highly paid, and evolutionary psychology predicts (and observation confirms) that women generally take fewer risks than men and so are less likely to compete for the highest-risk, highest-income tasks. If we want more women in politics and industry, which is a good thing precisely *because* they bring different values to bear, and if women are to have influence at the highest level, then we have to change the rules consciously so that influence is *not* linked simply to competitiveness, as is the case at present.

In short, if we want a society that is good for both women and men, and one in which female values are represented at least as much as male, then we should not pretend that women and men are identical creatures. We should recognize the differences and then create societies in which both sets of values are properly reflected. Otherwise we will again end up creating more traditional societies, in which male values prevail absolutely and/or in which women can succeed only by behaving like men. We can achieve this end not by denying biology, nor by making biology the basis of our ethics, but by recognizing our biology and then applying ethical principles to the facts of the case.

We should look more closely, too, as evolutionary psychologists do, at the essential prerequisite of mate selection, and all that goes with it.

CHOOSING MATES AND BEING CHOSEN

We see differences in reproductive strategy in the way that each individual—human and otherwise—chooses a mate. In very broad terms: The man's arithmetically ideal strategy is to find a quantity of women who are, above all, fertile, because he hopes to have as many offspring as possible. The woman's arithmetically ideal strategy is to find partners of the highest possible quality, to ensure that the relatively few offspring she is able to have acquire genes that are good partners for the female's own genes. Females who belong to species whose males are simply big and polygamous and do not take part in child rearing are advised to mate with the biggest, toughest males, whose big, tough genes will be passed on to the sons, who in turn will achieve enormous mating success. Females of species whose males do practice childcare—

including human beings, of course—are advised to find males who are good, steady, and will provide resources over long periods.

Unfortunately, cheating plays a large part in the affairs of all animals. DNA studies among birds that are apparently monogamous are increasingly showing that the chicks in any one nest have more than one father. An early study of dunnocks, or hedge sparrows, was among the first to show this, but more and more are coming on-line. By the same token, of course, the apparently devoted fathers have youngsters dotted here and there in more than one nest. In a sense, the policy suits both partners in any one relationship: each spreads its genes, but each ensures that youngsters are raised safely and that at least some of them belong to the attendant father. On the other hand, Matt Ridley points out that the arithmetically ideal strategy for women is to become pregnant by a rogue (defined here as a man who is good at attracting sexual partners) but ensure that the child is brought up by somebody nice (in traditional terms) who is prepared to be a loving husband and father. DNA studies have also shown that plenty of women (more than the investigators suspected) do follow the arithmetically ideal course. Many more women than is socially comfortable have taken their lead from hedge sparrows.

Given that the females *must* set such store by quality of mate, it follows that they must be as choosy as possible. If they are too choosy, they will not find a mate at all, but they ought to set their sights high. Male conceit over the centuries has decreed that men "choose" a wife—this is a common expression in literature—and of course they do. For many men, in many kinds of society, monogamy offers the surest strategy of reproduction, even if the arithmetic says otherwise. When men enter such a relationship they are, in effect, adopting a female reproductive strategy and must be as choosy as the females. But men need to be especially choosy only when they live in particular kinds of society: those that favor monogamy. Women, as female animals, should *always* be choosy. That is their inescapable default position. In general, then, throughout nature, the males compete for the females' attention—they lay out their wares and credentials—and the females choose. Despite the human predilection for monogamy, and other cultural overlays, the same does apply in humans. As a broad generalization, men display and women choose.

The first person to spell this out clearly was, inevitably, Charles Darwin. In *Origin of Species,* published in 1859, Darwin spelled out the idea of natural selection: the notion that lineages of creatures are shaped by their environment over time as they adapt to its vicissitudes. In 1871, in *The Descent of Man and Selection in Relation to Sex,* he described a second powerful force that shapes evolution: sexual selection. The notion here is that all creatures fail in the end unless they attract mates, so theory and common sense suggest that any creature *ought* to invest a very great deal indeed in features of anatomy, and in behavior, that will secure it a mate. The peacock's tail, Darwin pointed out, was inexplicable *except* as a device for attracting hens. It certainly did not enhance the cock's ability to survive day to day. It was, however, beautiful in the eyes of the hen, and it certainly attracted her attention. Twentieth-century studies have confirmed what Darwin took to be the case: the peacocks with the brightest tails (in general, those with the greatest number of "eyes") do indeed attract the most mates.

Nowadays Darwin's notion of sexual selection is one of the hottest topics in evolutionary research. It has launched if not a thousand, then at least a hundred research projects. Yet as Helena Cronin pointed out in *The Ant and the Peacock,* the idea was all but neglected for a hundred years after he published it. Apparently the late Victorians in particular felt that whereas *natural* selection was serious—since it seemed to focus primarily on the grim business of competition and survival—mate selection and choice were altogether too frivolous to be a serious shaping force. The fact that no sexually reproducing creature can pass on its genes *unless* it finds a mate seemed to escape them: mating is as deadly serious as eating or combat (and indeed involves combat). So, too, did the fact that peacocks demonstrably do have tails that clearly encumber them day to day, and that stags grow enormous antlers, requiring enormous outputs of energy, that they employ not to fight off wolves or to compete for grazing rights but purely to fight off rival males in the rutting season. *Of course* mating is serious, and so, therefore, are the evolutionary pressures that improve the chances of mating.

One biologist who did take sexual selection seriously was R. A. Fisher in the early decades of the twentieth century. He enhanced the notion with an idea known as Fisher's runaway. The truth is that a peacock's tail by itself will not enhance mating. It does the trick only if the peahens are psychologically attuned to respond to such a tail. So now ask how this cozy dialogue could have arisen in the first place: on the one hand, the wondrous tail of the cock, and on the other, the hen's predilection for such wonders.

Well, said Fisher, imagine that long ago, when peafowl still resembled chickens, one particular cock did indeed have a slightly longer tail than the average, and one particular hen was attracted by that tail. Imagine, too, that both the tail itself and the predilection for such a tail were heritable features, each underpinned by a hypothetical gene. The offspring of that particular cock and hen inherit *both* the gene that produces a big tail *and* the gene that produces the predilection for such tails. Of course, as we have noted in earlier chapters, genes are expressed only when they find themselves in an appropriate context; so we must hypothesize that the long-tail gene is expressed only when it finds itself in a cock, and the predilection gene is expressed only when it occurs in a hen. But there is nothing exceptional about such a suggestion. A great many genes that are carried by both sexes are expressed in only one of them, including genes that contribute to milk yield in mammals.

All we have to suggest finally is that the offspring of the original long-tailed cock and tail-loving hen thrive; indeed, that they thrive particularly well and produce many lusty chicks of their own. Those chicks, too, will inherit genes for long tails, and genes for a love of long tails. Soon we have a subrace of peafowl possessed with genes that code for long tails and the love of long tails.

So now the long-tailed cocks are surrounded by other long-tailed cocks and must compete with them for mates. The hens in turn, imbued with a love of long tails, must compete with others with the same preference. Among the cocks, those with the longest tails attract the most mates because they appeal most strongly to the hens' innate preference. Among the hens, those with the strongest predilections pursue the long-tailed types the most ardently. Thus, with each generation, both the long-tailedness and the preference for long tails are enhanced. Those in whom the tendency is strongest are always selected. This is Fisher's runaway. The proposal is that once the trend is established—a showy feature in one sex, and a preference for that feature in the other—it will become more and more exaggerated in each generation. All that will stop the trend in the end is natural selection. A peacock with a 20-foot-long tail would find it hard to survive at all. This example illustrates how natural selection and sexual selection can (and probably generally do) operate in opposition.

Why, though, should the peacocks with the showiest tails (and the hens that preferred them) have thrived in the first place? Why didn't the first pair, who indulged this fancy, simply die out? This is where the original idea needs some refinement. We need to suppose that the tail by which the peacock draws attention to himself in fact reveals—advertises—some serious quality that really does enhance survival. What could such an advantage be? What other quality—apart from its mere beauty—could the peacock's tail connote?

Actually, the beauty is largely the point. A cock that can grow such a tail has to be vigorous. It must be well nourished, showing that it has the ability to gather food. It must be free of parasites, for birds beset by worms and ticks are dull and mangy. A big, bright tail, in short, says to the hen, "I am a lusty, healthy bird! If it were not so, I could not grow this tail!" This, for a hen who wants to hitch her genes to the best mate, and give her offspring their best chance in life, is valuable information.

So the advertisement must reveal something real. But it requires one other quality if it is truly to be effective. It must *cost* something. After all, if it was easy to produce a big, bright tail—if it required no energy at all to make such a thing, and if parasites had no adverse effect—then the meanest, wormiest, stupidest peacocks could grow and parade enormous plumes, and the symbol would mean nothing at all. Hens who preferred big tails would find themselves mating with ne'er-do-wells, their offspring would suffer accordingly, and the lineage would die out. Fake symbols and a predilection for them quickly die out.

These simple and straightforward lines of thought lead us directly to one of the most extraordinary evolutionary insights of the late twentieth century: the handicap principle, first proposed by the Israeli biologist A. Zahavi in the 1970s.

Zahavi formalized the notion that in the first place, it pays animals to send signals to other animals about their state of health and well-being; and that in the second place, such signals will not serve their purpose unless they are real, or they will soon be disbelieved; and that finally, such signals will soon be seen through unless they are to some extent costly to the sender. A signal that costs the sender nothing is not worth the feathers it is printed on.

Put all these thoughts together, said Zahavi, and we see—astonishing though it may sound—that a creature that wants to signal something serious about itself ought to show the world (or some specific receiver) that it is able to incur some *handicap.* The signal itself should *be* the handicap. The peacock's tail says, "I am a big strong peacock" precisely because it handicaps its possessor. In effect, the complete message reads, "I can survive, despite the presence of tigers in the woods, and ticks and worms absolutely everywhere, and outsmart my fellow peacocks, *even though* I am carrying this enormous tail. In fact, I can spare a significant proportion of my diet to grow this tail even though I have to compete with predators, and parasites, and others of my own kind. What a fellow I must be!"

Because on the face of things Zahavi's idea seems so bizarre, some biologists were reluctant to accept it. But in the early 1980s, Alan Graffen at Oxford University quantified the idea—fed it into mathematical models—and showed that it worked perfectly well, and so the idea has caught on. Now, indeed, the handicap principle is seen to be one of the most powerful in modern evolutionary thinking.

Although we might risk the accusation of just-so stories, in the interests of brevity we can see how this idea applies to human beings. It lies, indeed, at the root of all conspicuous consumption. Those who sport gratuitous fur coats and huge, thirsty cars are saying, "Look! I am so rich I can afford to throw the stuff away—and still be richer than you, or anybody else, come to that!" The man who rushes to pay for the drinks is pursuing the same strategy: "Look! I can still afford the bus fare home even though I am squandering $40 on ten ghastly pints of undrinkable beer!"

We might even apply the handicap idea to the paintings of Anthony van Dyck presented in the summer of 1999 in a brilliant exhibition at London's Royal Academy. Van Dyck shows various beautiful young men from the court of Charles I in the 1630s and 1640s, with long curly wigs, lace trimmings, huge cuffs, and high-heeled boots with tops like big, floppy ice-cream cones. Their expressions are insolent, louche, and contemptuous. They seem the ultimate popinjays: self-indulgent and androgynous. *Decadent* is the common epithet of viewers; the young men's sexual proclivities the usual topic of discussion. And what of the painter's role? Does he approve of their corruption? Is he complicit? Or is he mocking them, exaggerating their conceit for the world to scorn?

Neither of the above, I think, is closest to the truth. In reality, the young men in the court of Charles I were warriors. They lived in dangerous times, and they knew it. They were seeking to defend Catholicism (whether they were devout or not is not the issue;

Catholicism was their cause) against what seemed an inexorable tide of Protestantism. They sought to defend the divine right of kings in an increasingly secular and republican age. The civil war was not far away, as they were well aware. They knew it would be their job to lead the armies and that they had a fair chance of losing. They were, however, good at what they did. Some at least were competent field officers, and all were accomplished swordsmen.

In truth, their effeteness was a double bluff: a handicap. It said, "We are *such* accomplished warriors, and so far above you in technique and intelligence, that we do not need to dress up as fighting men. Even when we are weighed down with feathers and trimmings and in danger of tripping over our own boots, we can beat you, or anything that those commoners might care to throw at us!" Such feigned effeteness, indeed, is a common strategy among fighting men. Dumas's three musketeers; British cavalry regiments of the mid–nineteenth century, whose soldiers feigned lisping, childish, and effeminate accents to throw their prowess into sharper relief; and many of the arch-villains of fiction, like Ian Fleming's Dr. No, with his soft furnishings and furry pussycats, have been examples of this tactic. It is also evident in many modern boxers. Some, like Mike Tyson, prefer always to look tough. Others, like Britain's Chris Eubank, desport themselves as dandies, as if butter wouldn't melt in their mouths.

Indeed, the chief problem with the handicap principle is that it seems to explain too much. It can be employed to explain just about every extravagance of human and animal behavior. Ideas that explain too much are innately suspect. It is for such reasons that evolutionary psychologists must rescue themselves from the charge of just-so stories and turn their grand speculations into testable hypotheses and then test them. They are aware of the need to do this, however. Many are among the sharpest thinkers in modern biology. Pointing out to them that they need to test hypotheses is like telling Olympic athletes they should tie their shoelaces before they attempt the 200-meter high hurdles.

The explanatory power of the sexual selection idea (refined by the handicap principle) has been startlingly demonstrated in the 1990s by Geoffrey Miller of University College London. He has argued that the enormous growth of the human brain between 2 million and 100,000 years ago—brain size more than doubled between the late *Australopithecus* and *Homo neanderthalensis* or the anatomically modern *Homo sapiens*—should be explained by sexual selection, and indeed that it can be explained *only* by sexual selection. Furthermore, he says, the things that human beings—particularly male human beings—do with their enormous brains are best explained as sexual attractants. Why did human beings develop the skills of dance? Not to impress the mammoths they were trying to kill or the bears they sought to avoid but to attract mates. Why develop painting to the level displayed by Van Dyck? Initially, to show off to potential mates. And so on and so on.

I think this idea is right but that there is more to it. The alternative, traditional view is surely right as well: that the human brain coevolved alongside the human hand. With the hand, the brain could turn its ingenuity into useful action. The presence of the brain provided selective pressure, encouraging greater dexterity. Greater dexterity in turn provided preconditions that would encourage greater brain power, to exploit the manual skill more fully. So as the generations passed, the two organs—hand and brain—egged each other on.

The traditional view explains the human brain in terms of survival selection, or straightforward natural selection. Perhaps Miller is right; perhaps this is not enough to explain why human beings also developed so many extravagances, from dance to fine art, which do not seem to enhance survival directly. Perhaps we need sexual selection as well, superimposed.

But this is the point. Commonly—as with the peacock's tail—natural selection and sexual selection work in opposition. The handicap principle effectively demands that this should be so. But in the case of the human brain, natural selection and sexual selection seem to work in harmony. Sexual selection does indeed prompt the brain and the body to work more subtly than mere survival seems to demand. The extra skills, however, though perhaps not strictly necessary, would certainly come in handy in hunting. By the same token, the skills that enabled a hunter to fashion a better arrow would also enable him to shape the jewelry that might win some Stone Age affection.

It seems worthwhile to stir in the third possible component: social selection. Human beings survive best when living in groups, and the bigger the group, and the more intelligent its members, the greater the social skill that is required to live sociably. About 60 percent of the human brain is occupied with social skill. But again, the ability that underpins sociality also, surely, underpins simple day-to-day survival and can enhance the refinements that win mates. In humans, then, it seems that survival selection, sexual selection, and social selection work together in ways that have not been the case in other lineages. This is why we have developed the way we have, while pigs, dogs, and elephants, bright though they certainly are, have not gone on to develop the tricks of civilization.

This, then, is a glance at one of the categories of notions that occupy evolutionary psychologists. Broadly speaking, it is concerned with "why" questions: why animals (including human beings) behave the way they do. The second great category is concerned primarily with "how" questions: notably, how the brain works.

HOW THE BRAIN WORKS; SPECIFIC QUIRKS AND PATHOLOGIES

Unsurprisingly, evolutionary psychology has attracted two main groups of thinkers: evolutionary biologists, and psychologists. The former have been mainly interested in behavioral strategies; the latter, in how the brain works.

The great shift has been from a view that sees the human brain as a giant all-purpose calculating machine, applying general rules to all problems, to one that sees the brain as a "modular" structure, with a variety of specific abilities only loosely grafted together. This latter idea has been summarized (primarily by John Tooby and Leda Cosmides) as the Swiss army knife model.

The modular idea seems to have begun with the attempt by Noam Chomsky, in the 1960s, to explain the human propensity for language. At that time the prevailing idea ("paradigm") in all psychology was that of the behaviorists, who suggested that people learn language by associating particular sounds with particular events. Chomsky pointed out that for various reasons such a model of learning simply could not account for the fact that all human beings learn their own local language more or less unerringly in the first few years of life. First, children seem to acquire far more insight into their own language than they seem to be exposed to. For example, nobody formally explains to small English-speaking children that to put a verb into the past tense you should add "-ed" to the end. Yet they learn this rule. Indeed, they apply it even when the local rules of the language say that it should not be applied. Small children at some stage invariably say "runned" instead of "ran." Adults see this as a mistake and chuckle indulgently. If we think about it, however, the child's brilliance is stunning. The child *infers* that the past tense of "run" ought to be "runned" from the fact that other verbs *do* follow such a rule. Indeed, the child invents the word *runned* without ever having heard it—since it is not used, except by other small children. Typically, small children first learn the "correct" form, "ran," parrot-fashion; *later,* as they infer the general rule (add "-ed" to make the past tense), they impose the "mistake," "runned."

Furthermore, our own ears tell us that different people speak different languages (there must have been many hundreds of thousands of different languages since human beings first evolved speech), and the textbooks tell us that different languages have different syntactical rules. Yet even cursory inspection shows the underlying similarity. In all languages, for example, there are nouns, verbs, and adjectives; and all, despite the many variations, have an essentially hierarchical structure, with subsidiary clauses nesting within main clauses.

The languages of human beings might be infinitely variable, but the ability to learn one's local language rapidly and generally unerringly, and with a minimum of clues, is a human universal, and beneath the surface variations the "deep structure" of all human languages (the syntax) is essentially the same. Language, in short, said Chomsky, is not just one more task to which the generalized, all-purpose brain can turn itself. It is a very special task indeed, which all human beings learn in a very special way. It is as if we all had a "language module": a specific part of the brain that is finely attuned to this particular task.

Cosmides and Tooby's Swiss army knife model applies this basic notion to virtually all the tasks human beings do. However, we cannot simply see the Swiss army knife as an extrapolation of Chomsky's language module for the paradoxical reason that Chomsky himself rejected the idea that the language module could have evolved by natural selection. He simply could not see how such a module could have evolved, and so denied that it had done so. But modern evolutionary psychologists take it as a premise that all the many modules of the brain (including the hypothetical language module) have indeed evolved. It may be difficult to see how, but it is less implausible to believe that the different modules have evolved than to believe that they have not. For how else does any living creature acquire its various attributes?

The module model is proving to have great power. Within the "language module" itself, for example, it now seems that the ability to spell, say, is coded by a different "submodule" or "subprogram" than the ability to infer grammatical rules (for example, that "-ed" must be added to verbs in English to put them into the past tense). Different people show different and highly specific defects—which are often proving to be heritable—in their ability to process language. At the same time, however, more and more psychologists doubt whether the different modules of the brain are really as separate as the blades and corkscrews in a Swiss army knife. Different parts of the brain do seem able to borrow abilities from other parts, and to adapt skills learned in one area to other areas. Brains seem to operate in ways that, in part, and at times, are reminiscent of the Swiss army knife, but at other times suggest that they operate as general, all-purpose calculators. There's a long way to go before we can claim to understand them.

This, then, is the briefest and cruelest possible outline of the main ideas in evolutionary psychology.

WHAT USE IS EVOLUTIONARY PSYCHOLOGY?

Since (I believe) evolutionary psychology offers worthwhile insights into human thought and attitudes, and since it is worthwhile to "know yourself," I am sure that it will prove, in the future, to be one of the most significant contributions to civilization. The general point is *not* that we *should* strive to behave in the ways in which we think we have evolved to behave. We must respect the naturalistic fallacy: we must, as a separate and independent exercise, contrive to develop systems of ethics that explore the fundamental ideas of fairness, justice, kindness, responsibility, guardianship, and all the other traditionally acknowledged components of good attitudes and behavior. Then we should adjust our behavior to those ethical principles, whatever ways our built-in proclivities may be urging us.

Evolutionary psychology provides three vital addenda, however. First, it can help us see when the thoughts and passions that we think are noble may in truth be no such thing, may in fact be mere expressions of some base, primitive drive. People in war experience extreme passions of patriotism, which often incite them to do things that in peacetime they would think horrible: set fire to villages; urge their sons and brothers to commit suicide in some foul trench. Perhaps if people could recognize the primitive roots of what they perceive to be their own nobility, they might take pause. What is "patriotism," except a kind of extended tribalism? And what is that but the "desire" of some gene to spread itself by an extreme preference for kin? Why is this "noble"? Why is it worth dying for, or sending others to their deaths? *Decorum est pro patria mori,* said Horace—"It is meet to die for one's country"—and many a young man has acted on this adage, of his own volition, or someone else's. An evolutionary psychologist would say, "Nonsense!" and, as a generalization, would surely be right.

Second, however—which may indeed be the opposite point— we may want to build societies that are fair, responsible, and all the rest, but we also want them to be *robust.* Many societies in the past have tried to *impose* a crude idea of what justice and responsibility are all about. Virtually all of them suppose that one particular group—the military, the aristocracy, the church, industry, the working class—should dominate the rest. Some civilizations powered by such rules have lasted for centuries at a time, although most, when we look back on them, seem to have been fairly horrible. Even the Greeks, whom we feel had some nobility, kept slaves and were constantly at war (city-state against city-state), and although they coined the term *democracy,* they never supposed that it should be extended beyond the circles of patrician males. But all societies, through all of history, have felt that they were "just" and "right," after their fashion. The unhappiness of most of the citizens was just the way of the world.

Surely we can do better than that. A society that is truly worth fighting for would be just (as defined by the best thoughts and feelings we can bring to bear) but would also conform to the real desires and aspirations of its people: a just society that was truly in line with our own psychology. This must be a worthwhile aim; perhaps, indeed, it is the ultimate aim of civilization, although we need not assume that "civilization" need be the ultimate aim of each individual human being. The idea that "human nature" is real, and that societies work best when they are in line with human nature, is certainly not new. Hegel, for instance, said something very similar. But we cannot achieve this unless we know ourselves.

Present-day laws provide a small example of the principle. In Britain (as in most countries), we have laws that forbid murder, and others that urge us not to drive our cars too quickly. On the whole we obey the first, but, at least when the road is clear, we ignore the second. Why? The point is not simply that murder is serious and speeding (per se) is not, although that is one way to put the matter. The point is that deep down, as if at the level of our genes, we *know* that murder is bad. Most of us know that if we murdered anybody, no matter how we might hate them, we would feel remorse for the rest of our lives. Even if we killed our worst enemy, we would probably spend much of the rest of our lives praying for forgiveness (or the secular equivalent thereof). But we have no such built-in antipathy to speeding. The opposite applies: we have a built-in predilection for thrills and risk.

Lawmakers recognize this distinction. They acknowledge that a bad law is one that cannot be enforced, no matter how reasonable it might seem; they acknowledge, too, that laws are especially hard to enforce if, deep down, people do not *feel* the point of them. In practice we don't need a law to forbid murder. We *know* that murder is bad, and the law merely provides formal endorsement. We do need a law to forbid speeding because otherwise we might well drive twice as quickly as we generally do. But even so, this law is widely flouted. Our society would surely be more stable if we framed laws—or, more to the point, social policy in general—that truly reflected the way we *are,* and not the way some lawmaker would want us to be. This is *not* to say, however (and this is where we invoke the naturalistic fallacy), that the way we *are* is the way it is good to be. But the principle does acknowledge that if we can devise laws and mores that really do conform to our nature, then they would be much more likely to succeed, and the resulting society much more likely to be tolerable and to endure.

The final general lesson of evolutionary psychology is that if we do allow our societies to go with the flow—to conform as far as possible to the way human beings really are—we would create much better societies than we have managed so far. For evolutionary psychology tells us that most people, most of the time, are "nice." We all of us behave badly some of the time—selfishly, aggressively, viciously—and some people behave that way most of the time. But most of us, most of the time, are fairly unselfish, reasonably kind, and not particularly pushy and have a great sense of fairness and justice and of the need to respect other people's dignity. The fact that most *societies* through history—including most nations—have behaved selfishly, aggressively, and often murderously is an unfortunate oddity that can be explained perfectly well by a little game theory. Perhaps when this fact, and its cause, are more widely recognized, we can do something about it.

The point is easily made by reference to one of the simplest models in game theory: hawks versus doves. The hawks and the doves are metaphorical, of course: they refer to individuals who behave selfishly and aggressively versus people whose mien is more passive, who never pick a fight, and who do not retaliate when provoked.

The following story can be told mathematically—ascribing numerical values, or scores, to perceived gains and losses—but it can equally well be understood intuitively. Suppose, first of all, we have a society that consists only of doves. These metaphorical doves can be human beings or ciphers in a computer game. For present purposes it doesn't matter. They are simply symbols of dovishness, which behave dovishly.

An all-dove society wastes no time at all on fighting. Where appropriate, every task is undertaken cooperatively, with all the gains that thus accrue. All energy is thus expended toward useful ends—or to sheer enjoyment—and all tasks are undertaken with high efficiency. The sum total of wealth and happiness within such an all-dove society is as high as it could possibly be. If all the wealth and happiness is shared equally, then each individual would do very well.

Game theory (and common sense) reveals, however, that such a society, though in some ways ideal, is unstable. For suppose, in the midst of such a society, a hawk appears. In genetic terms, we might say that a mutation occurs: a gene that hitherto encouraged cooperative peaceableness mutates into one that promotes aggression and self-centeredness. Such a lone hawk has an easy time. He grabs whatever is going, without working for it, and his dovish neighbors simply stand aside. The overall productivity, happiness, and well-being of the society go down because the hawk is taking without giving back. But the hawk himself is doing well. Indeed, he is doing better than any of the doves did before he arrived, when they simply shared everything equally. The overall cake is smaller than it was, but the hawk grabs the lion's share of it.

In short, an all-dove society, enviable and unimprovable though it may seem, is vulnerable. A hawk can come and prey on it.

So what happens next? The hawk grabs plenty of mates and produces plenty of offspring. Unfortunately—both for the hawk and for the society as a whole—the offspring inherit his hawkishness. Soon the society contains a great many hawks.

The circumstances have now changed. When the hawk was on his own, or when there were only a few hawks, they could swagger about taking what they wanted without fear of redress. Now there are plenty of hawks, and every now and again—in fact, more and more frequently—a swaggering hawk meets up with another hawk. Now, when the hawk simply demands, he finds himself in a fight. Soon there are fights everywhere. Everyone is now suffering. The fact that the overall efficiency of the group goes down is true but not relevant; the hawks at least are not thinking in terms of overall well-being. What bothers the hawks is that their unalloyed hawkishness no longer pays. They keep running into trouble. In a society shot through with hawks, the doves start to do better. They get involved in *fewer* fights (since they do not provoke any), and when they meet another dove, cooperation results.

So although the hawks do well at first, and multiply, the time soon arrives when they start to do badly. Too much hawkishness may indeed drive the society to extinction. What *cannot* happen is that the hawks will eventually produce an all-hawk society. An all-dove society is vulnerable, but as long as it remains uninvaded, it works very well. An all-hawk society cannot function at all.

It turns out, then, that the only society that is truly stable is one that contains both doves and hawks. The all-dove society is vulnerable to mutation—one that would turn a dove into a hawk. An all-hawk society fights itself to oblivion. We need not assume that any one individual within the mixed society is *always* a dove or *always* a hawk. More probably, any one individual will behave like a dove some of the time and a hawk some of the time, depending on circumstances and opportunity. At any one time, however, some will behave like doves and some like hawks. This is the only stable position, long-term; it is, in fact, what John Maynard Smith called the evolutionary stable strategy, or ESS. The idea of the ESS is one of the most important late-twentieth-century contributions to evolutionary theory, along with that of the selfish gene. Note that in the evolutionary stable state the society is *mixed.* It *cannot* consist only of hawks or only of doves. It must contain a mixture.

However, in an evolutionary stable society the doves seem bound to outnumber the hawks. But—and this is the snag—because the hawkish minority are aggressive, they rise to become the leaders. This, in a nutshell, seems to me to be the central dilemma of all civilization, not to say of all humankind. Most people are nice (where "nice" equals "dovish"), but nice people are bound to be ruled by nasty (equals "hawkish") people. Thus societies may contain a huge majority of sociable, hospitable, unselfish people and yet be ruled by leaders who are by all reasonable standards dangerously mad, and the kind, sociable, unselfish people follow their mad leaders into war and atrocity, just because they are easily led. Democracy ought to solve this problem. It ought to ensure that the majority of nice voters elect nice leaders. There are at least two drawbacks, however. One is that hawks can prevent democracy from happening at all (and the doves find it difficult to resist such hawkish ambitions); the other is that in reality, hawks put themselves forward as candidates in ostensibly democratic elections, while doves do not. Doves seem condemned by their dovishness to be overridden. How sad.

The overwhelming task for humanity, I suggest, is to devise systems of government that make it possible for doves to rise to the top, and, having risen, to conserve their dovishness. There is, clearly, a paradox here, so the problem is innately difficult. Always, however, the first requirement in solving problems is to define the nature of the task. This, I suggest, *is* the nature of the task: to overcome the central dilemma that is illustrated by the most elementary game theory.

This analysis surely explains why the world has had so few truly great leaders, or rather, so few of dovish mien. Of course, it has had many larger-than-life, hyperaggressive superman-style leaders: Alexander, Julius Caesar, Genghis Khan, Napoleon, classic hawks to a man. All had greatness of a kind, but all, when you boil them down, were fundamentally killers. The twentieth century suffered scores of such people, of greater or lesser magnitude: Stalin, Mussolini, Hitler, Franco, Antonio Salazar, Idi Amin, Saddam Hussein, Slobodan Milosevic, and so on, and so on. Great doves, however—leaders who preached and practiced cooperativeness and restraint—have been far more rare. From the twentieth century, Gandhi, Nelson Mandela, and the present Dalai Lama come most easily to mind. There have been others of this ilk, like Archbishop Tutu and Václav Havel, but it is difficult to extend the list much further. Yet *none* of the great doves has truly held power. All fought, throughout their lives, against some greater power. All in effect have been rebel leaders. We need not be cynical about this. We need not suppose that if Gandhi had ever had *real* power, he would have behaved like Stalin. Mandela was elected as undisputed president yet continued, magnificently, in dovish vein, helped by the redoubtable Tutu. It seems most likely, however, that *unless* Gandhi, Mandela, Havel, and the others had first been cast in the role of rebel leader, they would never have achieved power at all. In a straightforward run for power, they would surely have been outgunned and outsmarted by nature's hawks.

Jesus Christ was the archetypal dovish leader. He may have been somewhat fiercer in real life—more of an anti-Roman, fiercely pro-Jewish zealot. But the Jesus of the New Testament, as portrayed by Paul, is the apotheosis of the dove. As a political leader he is in the same category as Gandhi and the rest: leader of a group that was answerable to a more powerful oppressor. He also, of course, is the world's greatest and most unequivocal *advocate* of dovishness. Effectively, he says (in the vocabulary of modern game theory) that the prize of an all-dove society is *so* great that it is worth living with the short-term insults that are bound to accrue as the hawks begin, inevitably, to muscle in. An all-dove society, he says, is nothing less than "heaven on Earth." The insults of the hawks are a small price to pay for that. Besides, he says (which is perfectly good game theory), hawks can only get so far. Before long, when there are too many of them, they will see the error of their ways. This (so history tells us) is precisely what many Romans did. In fact, we could analyze Christ's philosophy formally: the attempt to push society from its evolutionary stable state—in which there are both doves and hawks—to one in which there are no hawks at all. Since the mixed society (hawks plus doves) is the stable state, to which societies subtend unless coaxed otherwise, the all-dove state requires obvious and constant effort. But that, nonetheless, is what Jesus asked us to aspire to. Game theory alone suggests that this is not a bad aspiration.

There are lesser and more immediate targets for evolutionary psychology. It is already beginning to influence medical practice, as some doctors begin to rethink their patients' "illnesses" in evolutionary terms. Much of the world's medicine has been devised to suppress fever. But evolutionary thinking—not in this case psychological, but never mind—says that fever is an adaptive response, evolved to suppress infection: quite simply, most pathogenic bacteria prefer low body temperatures to high. By suppressing fever, therefore, the doctor may be siding with the pathogen. Many psychiatrists, too, are beginning to take a longer, more biological view of mental "disorder." Grief and sadness are seen as evolved, adaptive responses: the mind's way of disengaging itself from relationships that once were taken very seriously indeed but have now come to an end. Depression is surely a pathology when taken to the extremes often seen in modern life—totally disabling. But pathologies in general are increasingly being seen as extreme manifestations of responses that, in more moderate form, are adaptive. Mild "depression" can be seen as the mind's attempt to take its possessor out of circulation, to regroup, a very useful tactic from time to time. Constant peak social activity or the attempt to maintain such activity is, almost beyond doubt, an inefficient and self-damaging strategy.

Critics of evolutionary psychology suggest that if we seek genetic bases for behavior, then we are ipso facto subscribing to "genetic determinism." Confusing the modern science with Victorian "social Darwinism," they then manage to conclude that a society that acknowledges the genetic roots of behavior is ipso facto red in tooth and claw and can have no respect for human individuality and freedom. (If these criticisms appear muddled when spelled out like this, and full of non sequiturs, that is because they are.) The critics thereby manage to conclude that the arguments from evolutionary psychology must be illiberal and inhumane, in contrast to those of the politically correct, which of course have everyone's best interests at heart.

The deep mistake in such thinking—and the *sin* of allowing ourselves to think simplistically, and merely following fashion—is nicely illustrated by contrasting approaches to the perceived "problem" of teenage pregnancy.

In 1998 he British government commissioned a report to analyze the problems of, and threats to, family life in modern Britain. There is a lot to be said for families, and a great deal of unhappiness in the world, so this in itself was surely a worthwhile thing to do.

However, the solutions that the government's committee suggested might politely, at best, be called banal; but the banality was culpable in its cruelty. The report took it to be self-evident that pregnancy among 15-year-olds was bad. It also observed (the data are undeniable) that such pregnancies occur mainly among the poor. Middle-class unmarried teenagers also become pregnant from time to time, but they are more inclined to resort to abortion than are poorer girls. In short, poor girls get pregnant more often and are more likely to continue with the pregnancies. Middle-class girls, if they get pregnant at all, are more likely to abort the pregnancy and begin again later, when they are married and have careers.

The "solution" to this unequivocal "problem" was perceived to be more "education." *Education,* however, in this context, meant more education in the crafts of contraception, since it was taken to be self-evident that young girls become pregnant primarily through "ignorance." The fact that children these days are taught about contraception when they are barely out of diapers was not allowed to influence the case. Clearly the girls needed more instruction or they would not behave that way.

At the London School of Economics, Helena Cronin and Oliver Curry applied a little evolutionary thinking instead. They pointed out that among *all* creatures—not just human beings, and not just animals, either, but plants as well—reproduction is a kind of last option. That is, for the first phase of their lives, *all* creatures concentrate on growth and on consolidating their position in life, their physical and mental strength, and their territory and resources. They reproduce when there is nothing more that can reasonably be done in the way of self-improvement.

This strategy makes perfect sense, whether analyzed mathematically or simply intuitively. All creatures have the greatest chance of reproducing successfully—of producing offspring that themselves have the wherewithal to reproduce—when they have consolidated their strength and resources. If they reproduce too early, before they are up to the task, they will lose out to those who take a longer view. But—and this is at least equally important—if they delay too long, wallowing in their strength, then they will be beaten to the gun by others who are less self-indulgent. For successful reproduction, in short, ripeness is all. Jump the gun, and you will fail. Delay too long, and you will miss out. If you survey all living creatures, you will find, unsurprisingly, that in all of them reproduction is beautifully timed. Some reproduce within weeks of birth (like mice), while others delay for the best part of a decade or even more (like eagles, dolphins, and elephants). But in all, the timing precisely suits their way of life. Almost always we can see why a significant change of timing would reduce their overall reproductive success.

So now apply such thinking to rich girls and poor girls. The middle-class 15-year-old readily perceives that she would be very well advised to wait until she is 30 or so before she has children. At 15, she is utterly dependent on parental good will. At 30, she will have a degree, a career, a salary, a well-heeled, stable partner, and also the option of complete independence should that partnership fail. She also perceives that between the ages of 15 and 30 life should be fun: university, holidays, parties, friends' weddings—the whole young adult experience. To get pregnant at 15 would be a terrible act of self-destruction. If pregnancy should ensue, then end it quick.

The poor girl has no such prospects. She will not be significantly better off at 30: no qualifications, no career, no salary, no stable partner. The 30-year-olds she knows—including, perhaps, her own mother—are no better off than she is but are already beginning to show signs of age (at least in the eyes of a 15-year-old). So the poor girl won't be any better off, but she will, she thinks, be losing sexual attractiveness. Furthermore, the passage from age 15 to 30 is not particularly enticing: more infinitely tedious school leading nowhere, while being treated like a child; more dead-end jobs; a string of pointless relationships. In short, there is nothing obvious to be gained from delaying reproduction. But there is perhaps a great deal to be lost from delay. Life in the short term would be *more* interesting as a young mother than as a schoolgirl and supermarket shelf filler, and in the long term she might herself be left on the shelf. Short-term motherhood is not at all bad, with grandmothers, great-grandmothers, aunts, and friends rallying to the cause, with herself as the center of attention. The middle-class "experts" who write government reports may see her pregnancy as a "problem," but she—very reasonably—sees it as a liberation.

So what is to be done? We could simply accept that for some people, childbirth at a young age is a reasonable option, and leave it at that. If we think it is not the ideal option, then the only humane solution is to help provide a way of life in which it really is in a woman's own interests to delay childbirth. "Education" always seems like a good thing. But the point is not to teach more contraceptive technique. The point is to improve life's prospects, so that a girl can perceive that at age 30 life really will be more comfortable than it is at age 15. Then there is a *reason* to delay, a reason other than that of pleasing the people who write reports. In *The Divided Self,* the psychologist R. D. Laing pointed out how difficult it is to see the world from other people's point of view, and how vital it is that we should try.

Finally we might note that eugenics has now become extremely unfashionable (absolutely *not* politically correct), whereas 100 years ago it was almost de rigueur in some intellectual circles. But the instinct that sees teenage pregnancy as a priori "wrong" is a eugenic instinct. If the generation interval is 15 years—that is, women reproduce at age 15—then in 100 years there are six generations. If the generation interval is 30, then in 100 years there are only three generations. Even if each of the women in both lineages has the same number of children, the lineage with the short generation time will still produce eight times as many children per 100 years as the lineage with the long generation time. Telling people to delay breeding, in short, is a way of telling them to have fewer children. I do not say it is *bad* to have fewer children. I do say that we should recognize eugenics when we see it.

I could give other examples. This alone reveals, though, that if evolutionary psychology is applied decorously, and all knowledge needs to be applied decorously, it could bring about significant improvements in our lives at all levels. At the individual level, it could improve our lot through better medicine; at the social level, through more enlightened social policy; and at the political level, by sharpening our insights into the nature of power. It will pay us to know ourselves.

Will we, though, seize these advantages?

THE FUTURE OF EVOLUTIONARY PSYCHOLOGY

There is, as we have already seen in this chapter, a huge amount of prejudice against evolutionary psychology. I have heard people declare (apparently with some pride) that they do not know what it is—and in the next breath condemn it, and all who take an interest in it. Expressions like "genetic determinism" and even "fascism" inevitably surface among the imprecations.

Two things in particular are needed if evolutionary psychology is indeed to advance and realize its potential. First, it needs better public relations. It needs to show that it has shaken off its equivocal roots, that it has moved a very long way indeed from the crude, nineteenth-century forays of social Darwinism. I like to argue that the ideas of evolutionary psychology should be seen to be Romantic, with a capital *R*. It is part of the Romantic ideal, after all—as powerful in Richard Wagner as in D. H. Lawrence—that human beings are "driven" by "forces" that well up from their past. The notion that these forces were originally shaped on the plains of Africa as our ancestors strove to survive and to come to terms with each other seems to me to be stirring indeed. The fact that these forces are encapsulated in lengths of DNA seems to me miraculous, and not innately threatening. On the other hand, the people who choose to reject the notions of evolutionary psychology out of hand—convinced that Darwinism equals fascism and that's an end of it—must attend to their own education.

On the other hand, evolutionary psychology must smarten its own act. Speculations must be refined into testable hypotheses; testable hypotheses must be tested, rigorously and over time. Serious practitioners of evolutionary psychology are well aware of this. Some extremely able students are now being attracted to the field, raring to go. All they really need is grants. But the world must be patient. Daly and Wilson's study of homicide, perhaps the greatest study in the field so far (and one of the most significant in all sociology), has taken more than 20 years. Perhaps it will be a good half century, and more, before society as a whole can and will draw confidently and as a matter of course on the insights of evolutionary psychology, as it frames social and political policy. But this time will surely come, as Mendel once said of his own work. All in all, the present strife surrounding evolutionary psychology is a pity. Of course we should continue to seek out the evolutionary roots of our own behavior, as this will surely prove to be one of the most worthwhile intellectual adventures of the next few decades.

It seems self-evident that our minds have been landscaped by our past, and that we should try to understand the constraints and predilections that our evolution has built into us. The presently fashionable view that our brains are a tabula rasa (to borrow the seventeenth-century expression of John Locke) is simply wrong, which is bad in itself. It has also proved to be cruel.

Yet evolutionary psychologists must beware of dogma, and of expressing too much certainty too soon. These are early days. Although human beings do have a distinctive "nature," which we should seek to understand, our behavior and emotional responses are obviously extremely flexible. Modern societies have only just begun seriously to indulge that flexibility, and only time will reveal the range of social possibilities that we and our descendents will devise and feel comfortable with. Overall, as always in human affairs, the point is to be humane. It is high-handed and potentially cruel to assume that people are infinitely flexible and to seek to impose some novel way of life on them, which effectively is what old-style Communists did in the Soviet Union. Yet it is surely humane to *indulge* the flexibility of our fellows and our children and allow them to explore their humanity to the full, in their own way. Imposition and indulgence are not the same thing at all.

In short, humanity must set itself two agendas. The first is to understand our nature more fully, a task in which evolutionary psychology is already playing a crucial role. The second, in this liberal and democratic century, is to explore as fully as we might the range of our own human possibilities. An optimist would suggest that both agendas are well in train. A pessimist sees strife, which is wrongly construed as a clash between science and ideology. With luck, though, the present strife will come out in the wash, and in a decade or less evolutionary psychology will take its proper place, not at the head of all attempts to understand ourselves, but as an acknowledged component of the broad endeavor.

EIGHT

Genes Rearranged

and Genes Conserved

"To breed" in the intransitive sense is to obey God's directive, in Genesis 1:28, to go forth and "multiply upon the earth." "To breed" in the transitive sense is either to cause other creatures to multiply or—and this of course is the meaning intended throughout this book—to manipulate their genes.

However, "breeding" in this sense—the manipulation of genes—also has two quite different connotations. One is what Mendel, Napp, Knight, and Bakewell did: take some wild plant or animal—or a domestic one that is not quite suited to the purpose in hand—and produce a creature that does precisely what a farmer (or a gardener, or a pigeon fancier, or a dowager duchess) requires of it. Wild creatures are adapted to the wild. The gene pools of wild plants include alleles that equip them with spikes or render them toxic. The gene pools of wild animals include alleles that make them swift, agile, lean, shy, or perhaps aggressive, willing and able to produce just as many eggs or babies and just as much milk as is compatible with survival and no more, because although nature as a whole is notoriously profligate, individual creatures cannot afford to be extravagant. Domestic creatures do what the farmer or the pigeon fancier or the duchess requires. The gene pool is stripped of untoward alleles: those that make spikes, and toxins, and prompt the baring of teeth. Mutations that cause anomalies—ten times more milk than is needed to raise a calf, or eggs by the basketful—are retained and encouraged. The gene pool as a whole is typically subdivided: not all the alleles are available to all the creatures. The subdivisions are commonly called varieties (in the case of plants) or breeds (in the case of animals), and each is roughly equivalent to the discrete subgroupings that occur naturally in the wild and are then known as races, or sometimes more grandly as subspecies. Wild plants and animals tend to be highly various; this is not always the case, but a certain degree of variety is part of the survival strategy. Domestic creatures tend to be uniform, or at least far more consistent. The farmer and the duchess—and the consumers—want to know what they are getting.

But "breeding" in the transitive sense also has a quite different connotation. At least in recent decades, breeding in zoos, reserves, and botanic gardens has become a serious and ever more vital arm of conservation. Of course, breeding in zoos and gardens does not replace conservation in the wild. Nobody ever said it does, or should. Increasingly, however, it is an essential backup. A population of wild tigers, for example, is unlikely to be viable in the long term unless it contains several hundred individuals because of the vicissitudes of the wild and because wild populations need to retain some genetic variability if they are to adjust in the future to changing conditions. But depending on how much prey there is, each individual tiger may require up to 100 square kilometers. So a reserve big enough to maintain a viable population of wild tigers may well need to stretch over tens of thousands of square kilometers. There are reserves for tigers in the wild, notably in India, but none is as big as that. India is a crowded country. So we can be reasonably sure that none of the present populations of wild tigers can be considered viable.

In time, however, things could get better. The human population seems likely to stabilize in the twenty-first century, and within a few centuries' time our numbers may start to diminish, not through coercion, but because people worldwide tend in general to have fewer children as they grow richer and more secure. That at least is the hope, and hope is a necessary component of conservation strategy. As populations decrease and agriculture becomes more efficient, it might be possible to re-create reserves that are big enough for tigers, and elephants, and the thousands of other creatures that are now endangered. But if we care about their long-term survival, then we have to keep them going during the next few difficult centuries, when there is too little room for them in the wild. In reality, too, wild populations are always liable to require some backup. This is the purpose of "conservation breeding."

However, the goals and hence the strategies of conservation breeding must be quite different from those of the farmer or the breeder of domestic dogs. Domestic breeders have an ideal in their heads, some vision of what the perfect apple, or cow, or poodle should look like, and as they move toward that ideal, they talk about "improvement." The conservation breeder has no such ideal in mind. The breeder of domestic apples or sheep deliberately reduces the gene pool of the wild creatures, retaining only those alleles that serve the commercial purpose. The conservation breeder, in absolute contrast, seeks to conserve *all* the alleles that are present in the original wild population, or, since this is liable to prove impossible, at least to retain the highest possible proportion. The term *improvement* can have no meaning in the context of conservation. The purpose of conservation breeding, however, is *not* simply to maintain creatures in their present state. The purpose is to retain enough genetic variety to enable a particular creature to resume its evolutionary path when, at some future date, it is returned to the wild. Contrary to what the critics often say, it *is* possible to return zoo-bred creatures successfully to the wild. This has not been done very often, because serious conservation breeding is a new craft. But it has rarely failed when carried out conscientiously, and no biologist ever expects 100 percent success. It's also of note that many *domestic* creatures—bred through many generations simply to be succulent or handsome and compliant—have taken wonderfully to the wild as soon as the cage door is open!

Very obviously, the breeding protocols required to reduce the variation in a gene pool, and to produce different subgroups that contain some of the species' alleles and not others, should be very different from those that are needed to retain all the original alleles. It is remarkable, however, how often the two kinds of enterprise tend to be conflated. Thus zoos in the past have sometimes tended to breed giraffes or tigers as if they were domestic animals, seeking to produce the tallest, the brightest, and so forth. The present worldwide fashion for white tigers is a manifestation of this. Such practices are highly pernicious. They are not exercises in conservation, although they sometimes pretend to be, and they waste space. Similarly, aviarists with cages full of rare parrots are sometimes mistaken for conservationists just because they produce six individuals where before there was only one. Well, there is safety in numbers. But if all the individuals in the flock are genetic replicas of each other, which is often virtually the case, and if the rarer alleles are allowed to disappear (which is highly likely), then again, such activities have nothing to do with conservation. So beware: do not assume that a "wildlife park" that produces an annual crop of baby animals to amuse the visitors is necessarily involved in serious conservation. It may be a serious institution, but babies per se do not make a conservation program. Neither, emphatically, does a zoo or a reserve that breeds antelope as if they were prize cattle contribute anything worthwhile.

It is worth looking briefly at what each of these two strategies of breeding really implies. Both are vital. We could not support a human population of 6 billion people without domestic crops and livestock, creatures that are often quite different from their wild ancestors. And a steadily increasing catalog of wild creatures will go extinct if we do not, in some measure, undertake conservation breeding. Furthermore, when we come to consider our own species—the past and present attempts in eugenics and the future possibilities of "designer babies"—we can learn much from those who deal with other creatures. As we will see in the next chapter, people who aspire to breed better human beings have a great deal to learn from breeders of millet and corn, although it never seems to have occurred to them that this might be the case.

BREEDING TO SPECIFICATION: THE CRAFT AND SCIENCE OF "IMPROVEMENT"

One problem in the "improvement" of crops and livestock is that the goalposts keep moving. In the eighteenth century British farmers wanted the fattest possible pigs. Eighteenth-century laborers working hard in God's fresh air needed plenty of calories, and fat is the most concentrated source. Traditional Yorkshire bacon was nearly all fat, and much prized (Yorkshire is a cold county). Even today, very thin slices of pig fat, known as lard, are featured on Italian menus, and very fine they are too. Eighteenth-century breeders accordingly imported Chinese pigs, which veer to the spherical, to cross with the lean, native Tamworths and others. Then, in the twentieth century, when people discovered nutritional theory but also began to forget what food once tasted like, leanness was again de rigueur. Modern bacon as approved by the European Community is all but uneatable, bulked out with water measured to the milligram: a legalized adulteration. But that is diversion.

The goals of domestic breeders are both aesthetic and functional: criteria that sometimes are in opposition but sometimes are complementary. Many an otherwise fine Berkshire pig has been turned into pork chops before its time because it lacked the four white feet that are favored in the breed, which is otherwise black. Perverse, you might think, because white feet have nothing obvious to do with performance. Yet the white feet showed, in an age before ear tags and computerized records, that a particular pig was indeed a Berkshire and not something else, and also had all the more practical, excellent features of that breed, including good mothering and a love for the outdoors, and a diet high in grass. The "white socks," apparently whimsical, were in practice a badge of authenticity.

Sometimes such badges detract from performance, however, at least in principle, though not necessarily in practice. It takes energy to produce horns that otherwise could be used to build muscle, so the long, droopy horns of Britain's traditional longhorn cattle were presumably costly, although the cost was not great when set against the massive bulk of the beast. Herefords, however, the great commercial beef breed of the nineteenth and early twentieth centuries (and still going strong) had stubby horns. The vast crescent horns of Africa's domestic Ankole cattle are almost pure aesthetics. Yet in context those adornments make perfect sense because cattle, in much of traditional Africa, serve as currency rather than as food (and as such they are highly convenient, if only because you don't have to carry them).

Nowadays, though, the breeding of livestock is generally about measurement: milk yield, rate of growth, number of eggs, and so on. Europe's farmers (as opposed to the professional breeders) began keeping records of pedigree around 1800. Now pedigrees that are routinely maintained by computer show who should be crossed with whom to maximize the chance of the perfect genetic combination. On the other hand, some animal breeding is almost pure whimsy, like the breeding of dogs specifically for show. The snooty, narrow-headed collies that win the rosettes are not, on the whole, the tough little packets of energy that round up sheep so efficiently in the cold mists of Wales.

Whatever the targets for improvement, breeders of crops and livestock adopt one or both of two main strategies.

When breeders (or farmers) are beginning with a population of wild creatures—or of domestic creatures that have not yet been "well bred"—the only sensible route at first is *mass selection*. The breeder simply decides what features are required, and then, with each generation, breeds again from that proportion of the population that meets the required criteria most closely. Thus it was that the earliest farmers developed (and subsistence farmers worldwide continue to develop) the informal varieties known as landraces. The farmer plants a field of, say, barley. Many of the plants subsequently die: some of drought and some of late frost, while a great many are killed by fungi and viruses. Some of the survivors are consumed, of course, but some are saved and replanted the following season. Generation by generation, by a process that resembles natural selection more than modern breeding, the farmer thereby selects a population of plants that is more and more closely adapted to local conditions. After all, the individuals that are not so well adapted simply die out, taking their alleles with them.

Typically, landraces vary from area to area, each one finely tuned to the particular slope of the ground, the pests that emanate from the local woods, and so on. Over time, the landraces begin to differ markedly from their wild ancestors, being, for example, less poisonous or more juicy. Typically, although each landrace will have a distinct character, the individual plants within it are highly various: some tall, some short, some flowering early, and some later. This variousness is helpful, in a primitive context: an invading pathogen or a late frost will kill some individuals but not all. "Monocultures," by contrast (where all the individual plants are genetically virtually identical), may be wiped out completely by any one epidemic. But variability has its drawbacks, too. Ideally, it would be good if *all* the plants in the field performed as well as the best ones do. Furthermore, although variation confers *some* protection against pests, such protection is rarely absolute. Effective, specific resistance (requiring specific resistance genes) is better. Subsistence farmers commonly lose at least half their crop to pests before it is consumed, meaning they must do twice as much work, and use twice as much land, as ought to be necessary.

Sometimes, then, landraces are the best option. But as conditions improve, and the farmer gains more control, more tightly defined varieties bred for particular qualities tend to prove superior. It is important not to jump the gun, however. Modern varieties need backup, such as pesticides, fertilizers, marketing, and bankers to underwrite the costs. The imposition of modern varieties in economies that are not geared up to them has sometimes proved very destructive.

At the genetic level, mass selection simply weeds out many of the alleles from the gene pool as the generations pass: the ones that lead to thorniness or sourness, or coarseness of fleece, or other undesirable characters. Soon (and mass selection can produce significant changes within a few seasons) the farmer qua breeder has produced a new gene pool that is much less various than that of the wild ancestors, with many of the polymorphisms reduced to "fixed" alleles. But as the seasons pass, more and more of the remaining alleles are of the kind that produce the characters that suit the grower.

Clearly, though, mass selection has limitations. In general it lacks precision. In addition, the breeder makes use *only* of those alleles that were present in the initial, wild gene pool. Often, extra qualities are required, and new genes—the livestock breeder would say new "blood"—must be brought in from elsewhere. Thus, English breeders of horses produced fine strong animals by mass selection but then introduced Arab blood for extra speed; and thus, too, England's eighteenth-century breeders of pigs turned to China for extra fat.

So we come to the second main strategy: *crossing and selecting.* In general, this works best when the parent lines are already well bred, with gene pools refined and narrowed to the task in hand, producing predictable phenotypes. The breeder begins with two parental "lines," or "strains" (or sometimes with full-blown varieties), that seem to have complementary qualities and crosses them. *Some* of the progeny will combine the good qualities of both parents. Some will combine the worst features of both. Most will be somewhere in between. Breeders of wheat in the British government's old Plant Breeding Institute at Cambridge used to make 800 or so different crosses every year—each one a marriage of different strains—to produce literally millions of offspring, of which just a few were then selected to be bred on to the next generation. Typically, after the initial cross, the desirable offspring would be allowed to reproduce, and then the best of their offspring would be selected. Then these offspring would reproduce again, and they were again selected, and so on, until, after about 12 generations, a new variety that was of high quality and "breeding true" was ready to be passed on to the farmers.

Each individual plant, of course, inherits half its genes from one parent, and the other half from the other. Often, however, breeders require *most* of the characters from one parent, and only one character from the other. Commonly, for example, they will cross a well-established variety (let's call it A), beautifully suited to its conditions, with some wild or primitive relative (which we'll call B) that just happens to contain some particular gene that confers extra resistance to some pest. The primitive parent confers its valuable allele but also donates thousands of others that are far less desirable. The breeders then undertake a series of "backcrosses." The first, or F1, generation contains 50 percent of alleles from A and 50 percent from B. The breeder then crosses F1 plants with A's. The consequent, F2 generation has now derived 75 percent of its genes from A and only 25 percent from B. Another backcross reduces the contribution of B to 12.5 percent. In each generation, the breeders ensure that the plants they retain are the ones that contain the required allele. If the allele in question is one that confers pest resistance, they can pick out the appropriate plants simply by exposing the whole lot to the pest in question and seeing which survive. The plants that remain after half a dozen or so backcrosses will contain almost no B genes at all, apart from the required resistance allele.

These general principles apply to both animals and plants. Plants, however, present an extra complication (although it is one that is also evident in less dramatic form in animals). As discussed in earlier chapters, most creatures benefit from some measure of "outbreeding." It is not wise to cross creatures that are too closely related, as there are specific benefits in heterosis (having different alleles on particular loci) and several possible drawbacks to too much homozygosity. Many a dynasty of plants and animals has been killed off by too much inbreeding.

However, some plants are specifically *adapted* to inbreeding. Often, in such "inbreeders," the female flowers are fertilized by pollen from the same plant, or indeed from the same flower. Mendel's garden peas were natural inbreeders, as we have seen: the sexual parts of the flower are entirely enclosed within petals, so that pollen neither leaves nor enters. Wheat and sorghum are also natural inbreeders. Biologists speculate that the ancestors of such plants were thinly scattered through the landscape, so that their chances of being fertilized at all were reduced: in such circumstances, it was better to forgo the advantages of heterosis and settle for pollen produced at home, so this was the strategy favored by natural selection. We must assume, however, that the gene pools of natural inbreeders do not include too many deleterious recessive alleles, which would otherwise cause trouble.

Other plants, however, are fully committed outbreeders. Some—including apples and plums—have "specific mating barriers," which prevent pollen from the same plant, or from genetically similar plants, from fertilizing the flowers. Thus we find that in general, two apples of the same variety will not fertilize each other. Since fruit are produced only when the ovum is fertilized (except in plants like domestic bananas that produce fruit parthenogenetically), growers get no fruit at all unless they raise compatible varieties within a bee's-flight distance of each other. It is bizarre that two creatures of the same species cannot breed together, but in plants this is often the case. Other natural outbreeders do not have specific mating barriers, but they will suffer severely nonetheless from excess homozygosity. Corn and millet are of this kind, and so are many pulses.

Consumers like crops that are reasonably uniform: they like to know what they are buying. Modern farmers also prefer reasonable uniformity, first because they want to please the consumers, but also because they want to be sure that all the plants will respond equally and eagerly to the extra-expensive fertilizers they apply to the fields and that they will all be ready for harvest at the same time, whether by machine or by gangs of hired labor. Breeders, in their turn, have to supply the farmers with seed that is correspondingly uniform. They also have to supply that seed in vast amounts.

When the plants are natural inbreeders, like wheat, it is relatively easy to produce uniform seed in large amounts. Natural inbreeders do not suffer excessively from homozygosity. So the breeder could, in principle, start with just one—highly homozygous—parent plant and multiply it by self-pollination, and all the offspring would be genetically the same as the parent and as each other, or at least very similar. The offspring would be highly homozygous, but as inbreeders they would not be unduly affected by this.

However, when the species in question is a natural outbreeder, simple multiplication of a single self-fertilized parent is not an option. Natural outbreeders *need* to be heterozygous. But heterozygous plants do not breed true for all the reasons discussed in earlier chapters. How, then, can the breeder reconcile the two requirements of high performance and uniformity?

In many crops, asexual reproduction comes to the rescue. The breeder of potatoes may produce an excellent plant by the normal methods of mass selection, followed by crossing and selection and then simply multiplying that plant by tubers. The tubers are all clones, genetically identical to each other and to the parent. Similarly, fruit breeders multiply their favored varieties by cuttings.

Thus it is that all the Cox's orange pippin trees in the world have been cloned by cuttings from the first Cox, which was bred by normal sexual means in the nineteenth century. Each of the many different varieties of grape, from aglianico to zinfandel, is also a clone. Each vine plant or apple tree may be highly heterozygous, yet all the different plants of any one variety are genetically more or less identical.

The cereals and pulses, however, which are the world's most important crops, cannot be multiplied asexually. They reproduce only by seed. When the cereal or pulse is an outbreeder, it becomes very difficult indeed to reconcile the two opposing demands: for uniformity on the one hand (which the consumer and farmer require), and for heterozygosity (without which the crop itself will suffer) on the other.

There are two main ways around this problem. The first is to produce a population of plants (a) whose gene pool is as purged as is possible of frankly deleterious alleles, (b) which contains just enough polymorphism to ensure reasonable heterozygosity in each plant, and (c) in which the different alleles that may occupy any one locus are indeed different yet produce much the same phenotypic effect. So two individuals in the same variety may look and behave much the same but may nonetheless contain different alleles. At the molecular level—which is where it counts—the need for heterozygosity is thus satisfied, yet the phenotypes are reasonably uniform. Overall, however, the approach is essentially one of compromise. The different traditional varieties of scarlet runner beans, for example, have gene pools of this kind.

The second strategy is to produce first-generation—F1— hybrids. Each of the parents of the F1 hybrid may be fairly homozygous, and therefore, since the plant is an outbreeder, may well look rather feeble. But when the two are mated, the resulting first-generation offspring are highly heterozygous; if the parents are themselves well bred, despite being phenotypically feeble, and are well matched, the F1 hybrids could be very vigorous indeed.

Such F1 hybrids are highly heterozygous *and yet* are uniform. Of course, F1 hybrids do not themselves breed true, so farmers and growers cannot simply save F1 hybrid seed and hope to get the same results again. They must buy fresh F1 seed each year. For commercial growers in developed economies, however (or indeed for weekend gardeners), the F1 hybrid is an excellent option. More and more crops are grown as F1 hybrids. It transpires, too, that even inveterate inbreeders do benefit, after all, from some heterozygosity: so even wheat, these days, is increasingly grown in hybrid form.

Livestock farmers play comparable games. British dairy farmers typically raise Holstein cows, which give plenty of milk, but cross the cows with a bull of a beef breed—traditionally with a Hereford but increasingly, these days, with one of the big European breeds like Charolais—to produce an extra-beefy calf. Britain's sheep farmers ring endlessly complex changes, commonly keeping F1 hybrid ewes on the hills (generally of medium body size) that they cross with big meaty rams (such as Suffolk or Leicester) to produce fast-growing "triple cross" lambs that are then fattened in lowland pastures.

For my part, I just love agriculture, at least in its more traditional forms, before so much of the countryside became a sprawling, rural factory. In all parts of the world it is good to look into fields and ask what exactly is going on. No cow ever stood on a hillside except for a *reason*. Why that particular breed in that particular place? There are many pleasant surprises in modern Britain, away from the prairies and the big sheds, where more and more hobby farmers are raising Highland cattle that look like musk ox, or Belted Galloways, jet black with broad white cummerbunds.

If you don't share this obsession, you may be asking, "So what?" Well, the key theme of this book is to explore current ideas, including the notion that human beings might be "improved" genetically, at first by "eugenics," and in the future by genetic engineering. I will discuss this further in chapter 9. Here, though, I ask you simply to ponder the problem of producing a "better" human being—more clever, more handsome, more athletic—in the light of the above comments about wheat and cattle. Wheat breeders produce *millions* of offspring to get the right combinations. The millions that do not come up to scratch are simply thrown out. Breeders of corn know they must conserve heterozygosity. It seems to me almost beyond question—even in the absence, as yet, of any firm data—that human qualities such as intelligence, insofar as they are influenced by the genes, *must* operate best when they are highly heterozygous. For good genetic reasons, a great corn plant may be the heterozygous offspring of feeble homozygous parents, and it is the job of the breeder to recognize the genetic merits of the parents beneath the unprepossessing phenotype. Similarly, we might commonly expect human geniuses to arise out of the blue, as the genes of each rather ordinary parent click in some magical but never to be repeated combination. Beethoven and Newton, both of irredeemably ordinary birth, seem to illustrate that principle perfectly, though Mendel's parents were obviously intelligent. Some measure of mass selection might well produce significant changes in human dynasties within a few generations, although of course each generation in humans lasts a great deal longer than in wheat—25 years as opposed to one year, or six months if you use greenhouses. But Adolf Hitler's ambition to breed better people by crossing and selecting can be seen immediately to be ludicrous. Science cannot tell us what is right, ethically and politically, but it can tell us what is feasible. Hitler's particular eugenic balloon can be pricked simply by reference to scarlet runner beans. Oddly, that was just as true in the 1930s as it is now.

Breeders of plants employ other tricks, too, which are not available to breeders of animals. We have discussed one of them: the creation of brand-new species by crossing different types, followed by polyploidy. The rutabaga arose as a polyploid hybrid of cabbage and turnip. We have seen, too, how August Weismann finally put an end to Lamarck's idea of the inheritance of acquired characteristics. He pointed out in the late nineteenth century that the germ cells, which produce the gametes, develop quite differently from the somatic (body) cells, so mutations or other changes in the latter should not affect the former. However, the same does not apply in plants. In many plants, virtually any tissue might redifferentiate to produce flowers and hence seeds. So a mutation almost anywhere in the plant might find itself in a seed and be passed on to the next generation. Breeders make use of this phenomenon in many ways. One is to grow plant tissues in culture, where some mutation inevitably takes place, leading to what is called somaclonal variation. Then entire new plants may be generated from the mutated tissue to produce entire new lineages. Modern breeders of crops also induce a great many mutations artificially in their search for new variations.

However, the details of breeding would require an entire book. We should move to the art and science of breeding for conservation purposes. In almost all respects, the strategies are quite different.

BREEDING FOR CONSERVATION

The world's first breeders of dogs—beginning perhaps 100,000 years ago—began with the gene pool of the wolf, and, after many a summer, their successors have produced a catalog of show dogs that extends from the Pomeranian and the Chihuahua to the mastiff and the Leonburger, with dachshunds, Dandy Dinmonts, and German shepherds in between. It seems an astonishingly creative process that so many varied creatures should emanate from ancestors that all seemed much of a muchness.

But modern dogs, astonishingly various though they are in phenotype, contain very few alleles that are not present in wolves—that is, that were not present in the wolves of 100,000 years ago. A few novelties must have been added to the gene pool by mutation, but for the most part each modern breed merely teases out, and emphasizes, a small proportion of the alleles that made up the gene pool of the pristine wolf. In short, whenever you breed to a prescription—whether for show dogs, dairy cattle, or field crops—you must *narrow* the initial gene pool. Individuals produced along the way that do not perform as the breeder requires are simply cast aside, and any esoteric alleles that they may contain, however valuable they might have proved in the wild, are thrown out with them.

Conservation breeders, on the other hand, aim to retain *all* the alleles present in the initial gene pool. In practice, compromise is necessary, for it is rarely possible to conserve every one of the initial alleles. Serious modern zoos that take part in cooperative breeding programs have widely agreed on an arbitrary but sensible goal: to conserve 90 percent of the initial genetic variation within each of the species in their care for 200 years. In 200 years, after all, it might be possible to re-create safe environments in the wild for the animals to return to; either that, or the zoos of the twenty-third century will simply continue the programs. The protocols adopted by the conservation breeder are absolutely different from those followed by breeders who seek to "improve" livestock and crops. When conservation breeders adopt the techniques of the livestock breeders—as they sometimes have in the past and may still, inexcusably, do today—then their animals are in deep trouble.

Conservation breeding is an exercise in population genetics. The particular qualities of individuals are not of prime importance: what matters is the total allelic variation within the gene pool as a whole. Some quirks of population genetics are very much against the conservation breeder, but others are serendipitous. In breeding for conservation, there are a few unexpected sources of happiness.

The great enemy of conservation breeding is *genetic drift*. Small populations have an inexorable tendency to lose alleles. So do big populations. But the likelihood of loss is greater in small populations, and there is far less opportunity to acquire new alleles by mutation because there are fewer individuals in which those mutations can take place. There are various reasons for this. One is that some animals in any generation will die without producing offspring, and their genes die with them. If alleles are rare, then, by definition, they are contained within only a few individuals. If the population is small, then some at least of the rarest alleles may well be contained within only one individual. If he or she dies without issue, then that allele is gone from the population forever.

But there is a more constant cause of loss by genetic drift. Each individual passes on only *half* of his or her alleles to each offspring. If the individual belongs to a species that habitually produces huge numbers of offspring in the course of a lifetime—a fly may have thousands, for example—then there is a very good chance that each individual will pass on all of its alleles to its collective offspring. But if the animal is a rhinoceros, or an orangutan, and normally produces only half a dozen offspring in the course of its life, then there is a fair chance that at least some of its alleles will not get passed on at all. If the total population is large, then any one allele ought to be contained within more than one individual and there is a very good chance that among them the breeding animals will pass on all the alleles that they contain. But if the population is small, and the rarer alleles are contained within only a few individuals, then again, there is a good chance that some of the rare alleles will be lost as each generation reproduces.

The rate of loss by genetic drift increases as the population goes down, and decreases as the population goes up; so to conserve genes, the biggest possible population must be maintained. On the other hand, alleles are lost by drift only when the animals reproduce. So the rate of loss by drift *in a given interval of time* depends on the number of generations within that time. The longer the interval between generations, the lower the rate of loss by drift. The generation interval of a mouse may be as little as nine months, so in our arbitrary but sensible period of 200 years, it will fit in more than 250 generations. But an elephant may have an average generation interval of around 30 years, fitting in only six generations in 200 years.

Thus loss by genetic drift is less in big populations, and in animals with long generation intervals; and greater in small populations and in animals with short generation intervals. It follows that *the shorter the generation interval, the bigger the population of animals must be to minimize genetic drift.* Now if we do some math (the details of which need not delay us, though they are standard fare to conservation geneticists), we find that we would need to keep a breeding population of more than 1,000 mice if we are to conserve 90 percent of the allelic variation over 200 years. But we could retain the same percentage of allelic variation over the same period with as few as 35 or so breeding elephants.

That, I feel, is an astonishing thought: that with as few as 35 breeding elephants, we could retain 90 percent of the allelic variation of the founding population for 200 years. It is also encouraging. Many countries—especially if they worked cooperatively—have the wherewithal to maintain a breeding herd of 35 elephants, and any one such effort could effectively save the species. However, life is not quite so simple. The 35 elephants would all have to be ready and able to breed, but a real herd of elephants would contain some that were too old, some that were still youngsters, and some that were simply not required for breeding purposes, because they might be genetic duplicates of others. The 35 required for breeding are called the effective population, and in practice at least 100 elephants would be required to maintain an effective population of around 35. Still, 100 is not many. This, then, strikes me as serendipity.

There is more serendipity. It happens that two highly heterozygous individuals, male and female, chosen from different parts of an animal's range, could between them typically contain around 75 percent of the total allelic variation of the wild population. Six well-chosen individuals should easily contain more than

95 percent of the total allelic variation of the wild population. These figures apply in a rough-and-ready way *whatever* the population. So if we began a herd of Asian elephants in Europe with six "founders" well chosen from the wild, then that herd could in principle contain 95 percent of the total allelic variation of the entire species.

Even when they get the numbers right, however, and begin their herds with well-chosen founders, conservationists still need to adopt fairly strict breeding protocols to maximize the proportion of alleles that are passed from generation to generation. The sex ratio matters, for example. A breeder of prize cattle might happily keep one stud bull with, say, 20 cows. This is fine for farming purposes. The bull would be an outstanding animal, and all the calves would partake of his fine qualities. But the genetic variation of the offspring would clearly be limited, since every one of them would obtain half its genes from the same parent. A conservation breeder, interested in maximum diversity, would clearly need to keep more than one male. In fact, it transpires—the math proves this, but the point can be seen intuitively—that diversity is maximized when the sex ratio is equal—the same number of males as females. Ideally, in any one generation, all the males would mate with all the females. This is rarely practical, and not vital, but equalization of sex ratios certainly does matter.

This can raise all kinds of practical problems, however. In the case of elephants, the males are very big and often dangerous. An effective population of 36 elephants would be a handful if 18 of them were bulls. Bulls can be kept safely and with kindness, but it takes a lot of money to provide safe, adequate quarters. With gorillas and many other animals, one male tends to dominate the breeding even when there are several males. The females prefer to mate with the dominant (alpha) male. It is very difficult to arrange all the genetically desirable matings without severely overriding the animals' natural behaviors. In such cases, judicious sterilization of the dominant animal and artificial insemination can help; again, this is difficult in practice, and some curators find such interventions distasteful.

It is important, too, that each individual should pass on the same proportion of its alleles to the next generation as all the others. So family sizes should be equalized. But in any one population, some individuals will be more prolific than others. Breeders might feel that this simply shows natural selection in action and that they *should* continue breeding from the animals that breed best. Again, though, this would be a mistake. Among other things, the animals that breed best in one situation (in captivity, say) may not be the ones that would fare best in another situation (back in the wild, perhaps). The idea must always be to minimize the loss of variation by drift, and the theory shows that this is best achieved by ensuring that each individual has the same number of offspring. But this means that if one female tiger produces five cubs, and another produces two, then only two of the cubs from the more prolific family should be allowed to breed. In real life, where space is limited, the three cubs not required for breeding might have to be culled. It seems perverse to kill individuals in what is supposed to be a last-ditch breeding program, but simple theory says that this is necessary if we really want to conserve animals over the long term. Again, though, many curators find such requirements distasteful.

All in all, then, conservation breeding is not easy. I discussed the ins and outs in an earlier book, *Last Animals at the Zoo,* and although the examples have changed since it was published in 1991, the principles are the same.

All the above comments seem to apply to animals bred in zoos. But surely it is better to conserve them in the wild? The answer is that in principle, *of course* the wild is preferable. But for some animals, as things are at the moment, conservation in the wild is simply not sufficient. We have noted one reason: that for some species, such as tigers, wild reserves simply cannot be made large enough to accommodate large populations. To be sure, with slow-breeding animals—which in general means big animals—it is possible to maintain a high degree of genetic diversity even in remarkably small populations. But it is also clear that this can be done only when the animals' breeding is orchestrated. Wild animals make their own mating arrangements. Among wild gorillas and black rhinos, a few males dominate the breeding. With wild animals, too, some individuals really will breed more prolifically than others, and there are no curators around to equalize the numbers. So when populations grow small among wild animals, the rate of loss by drift can be horrendous.

If the population remains small for several generations, the remaining animals will be almost homozygous. Populations that are greatly reduced—and stay reduced for several generations—are said to pass through a *genetic bottleneck*. Modern-day populations of cheetahs in Africa, elephant seals worldwide, and lions on the Serengeti are extremely homozygous, indicating that at some time in the past each of them passed through a bottleneck and lost allelic variation by drift. All of those creatures are lucky to have survived, because extreme homozygosity often leads to extinction through inbreeding depression. They remain vulnerable. All the cheetahs of Africa would be roughly equally vulnerable to the same pathogens, and one epidemic could, in principle, wipe them out.

Breeding in captivity does not, however, *necessarily* contribute significantly to the survival of species. A captive population of 50 parrots could well maintain a very significant proportion of the total genetic diversity of the particular species (given that parrots can have a long generation interval) if the founders were well chosen, the sex ratios and the family sizes were equalized, and so on. But 50 parrots who are all descendants of the same founding pair and are being allowed to breed randomly—or are being selected for the brightest feathers, or the shiniest beaks, or some other character, as if they were prize chickens—would probably contribute nothing at all worthwhile, especially if the flock was isolated and not exchanging genes with other flocks elsewhere. In short, the breeding of exotics is not necessarily conservation, and if the founders have been caught from the wild, as still may be the case, then it can be the opposite of conservation. On the other hand, if a good zoo takes animals from a wild forest that is destined to be devastated in the ensuing months by a timber concession and integrates them into a serious breeding program, then that is conservation. It is important to know the difference.

Finally, many people who are interested in wildlife conservation see it as essentially a "romantic" and certainly as a spiritual exercise. So it is. The economic arguments in favor of wildlife conservation are necessary but not sufficient. Take away the romance of it, and the emotion, and it loses its point. Unfortunately, many people who feel the appropriate emotions tend to see high technology as the natural enemy of spirituality. However, several of the modern biotechnologies—including cloning—have a great deal to offer the conservationist. Conservationists should not eschew high tech. As General William Booth rhetorically put the matter when he founded his Salvation Army: "Why should the Devil have all the best tunes?"

But it is time to move beyond the ideas and methods of classical genetics and breeding protocols, to discuss the family of new technologies centered around genetic engineering.

THE ROOTS OF GENETIC ENGINEERING

Many modern commentators are fearful of genetic engineering. The power it offers, they feel, is just too great. It frightens people at the practical level, with threats of novel life-forms out of control. More broadly, critics often argue that transgenesis is "unnatural"; it therefore raises the specters of blasphemy and hubris, which go far beyond mere practicalities. Advocates of genetic engineering reply that such misgivings are muddleheaded. Any day-to-day problems can be contained, and on the broader front, genetic engineering is no more "unnatural" than the ancient crafts of breeding, which farmers and dealers in horseflesh have practiced for many centuries.

For my part, I am sure that we should continue with the science and technology of genetic engineering. It has so much to offer on so many fronts: insights into how life works; the treatment of human disorders; clean, neat agriculture; wildlife conservation. But the potential is indeed enormous, and the criticisms must be taken seriously.

The argument that says that *all* interventions are "unnatural," however, and that genetic engineering "merely" extends traditional crafts, is at best disingenuous. There is a brief but cogent catalog of qualitative differences between the changes wrought by genetic engineers and those made by traditional breeders.

First, and crucially, traditional breeders in general can make use *only* of those genes that are within the sexual compass of the organisms. In general, animals or plants can be crossbred only with others of their own species: corn with corn, mice with mice, and that's all. In fact, there are many exceptions. Modern breeders induce novel mutations with the aid of mutagens and thus create alleles that have never existed before. By various maneuvers, plant breeders in particular have organized many a cross between different species. Plants form new species from interspecific crosses by polyploidy. Modern bread wheats are hexaploid, meaning, in effect, that they contain more chromosomes than they need, which enables breeders to produce "lines" that lack one or more chromosomes, or have one too many. By judicious crosses breeders can shift whole chromosomes from one lineage to another. Viruses, too, transmit genes between organisms that may be quite unrelated. The extent to which viruses bring about such "horizontal transmission" is unknown, but it may be significant. Such transmission, however, is beyond the direct control of the traditional breeder. In general, the gene pools that traditional breeders work with are circumscribed within more or less impregnable biological boundaries.

The genetic engineer, by contrast, can in principle take genes from any organism and put them into any other: human into sheep, fungus into mouse, cabbage into human. Charles Darwin proposed that all the creatures now on Earth descended from the same common ancestor and that we all, therefore, belong to the same vast family tree. With the aid of genetic engineering—and not otherwise—all organisms now on Earth belong to the same gene pool. A mushroom, potentially, becomes a fit mate for a human being.

Second, a traditional breeder who crosses two creatures to combine the desirable features of both in fact creates a hybrid that contains 50 percent of one parent and 50 percent of the other. Any one offspring may or may not contain the good features of either or both parents, and is at least equally likely to combine all the bad features. Genetic engineers, in principle, can simply introduce the required genes ad hoc. If they want to introduce a disease-resistance gene from some wild grass into a cereal, they do not have to make an unhappy hybrid and then undertake half a dozen generations of backcrossing to dilute out all the bad features of the grass—the low yield, the inability to respond to fertilizer, and all the rest. They can simply pick out the required resistance gene and, in principle, pop it in. In time they might create crops like Christmas trees. The species will hardly matter. The initial plant will merely provide a framework on which desired features are hung like parcels, to the growers' orders.

Yet this is not the limit of what the engineers can do. They can remove genes, too. Or they can take them out, alter them, and put them back. In principle (and increasingly in practice), they can make fresh genes from scratch. They can arrange the nucleotides in the order required to produce a protein to a precise prescription, and protein chemists are increasingly able to predict how a given protein will fold (that is, they can predict the tertiary structure that will result from a known primary structure), and hence to predict how that protein will behave. Through genetic engineering, the ancient craft of breeding is meeting the modern science of pharmacology. We have the basic techniques required to reconstruct and redirect the fundamental mechanisms of metabolism, even if, at present, we lack the knowledge to apply these techniques in more than a few simple contexts.

In short, traditional breeders—albeit helped by serendipitous tricks of nature, such as plants' propensity for polyploidy—operate within strict biological boundaries, which in general are the reproductive barriers that divide each species from all the others. By contrast, the genetic engineer need acknowledge no constraints at all except those imposed by the laws of physics. That, truly, *is* a qualitative shift. It is much better to recognize it as such, and to frame laws and mores accordingly, than to obscure the distinction. In the long run, fudging is never a good idea.

"Genetic engineering" became a reality in the 1970s, but the first stirrings can be traced to the 1950s, even before Crick and Watson had worked out the 3-D structure of DNA. At that time, the Italian-American biologist Salvador Luria showed that bacteria could ward off invading phage viruses by chopping their DNA into pieces with the aid of enzymes. These enzymes were called restriction enzymes because they restricted the viruses' attack (though this does not seem a particularly appropriate name now that their modus operandi is better understood). Then, in 1970, Hamilton Smith at Johns Hopkins University isolated one of these restriction enzymes from a bacterium, *Hemophilus influenzae.* This particular enzyme attacked not phage viruses but another bacterium: the ubiquitous *Escherichia coli. H. influenzae* escaped attack by its own restriction enzymes by protecting the vulnerable points of its DNA with methyl radicals.

The *H. influenzae* restriction enzyme, it turned out, attacked the *E. coli* DNA only at specific points. That is, the enzyme recognized particular sequences of bases. When it perceived such sequences, it latched onto them and cut the DNA. The particular sequence that attracted this particular enzyme was G, T, any pyrimidine, any purine, then A, then C: six in all. The bits of DNA that result after a restriction enzyme has had its way are called restriction fragments. Clearly, each macromolecule will yield its own particular pattern of restriction fragments when exposed to a particular restriction enzyme, and since the fragments can be separated by various means (in general involving migration through an electric field), they provide a "signature" for any one kind of DNA.

Nowadays, hundreds of restriction enzymes are known, produced by many different organisms. Each has its own preferred site of attack. Some latch onto particular sequences of four bases, some onto five-base sequences, and some onto eight-base sequences. Clearly, there is more chance of encountering a particular four-base sequence than a particular six-base sequence, so those that attack four-base sequences produce smaller restriction fragments than those that attack six-base sequences. When any one kind of DNA is exposed to a variety of restriction enzymes, one at a time, it yields a series of restriction fragments that start and stop in different places and that overlap. The sequence of bases in each fragment can be analyzed, and by matching the overlaps it becomes possible to construct a complete base-by-base sequence of an entire DNA macromolecule: in fact, of an entire chromosome. Repeat the process chromosome-by-chromosome and you end up with a complete printout of the entire genome. This is the basis of *genomics:* the endeavor first to map all the genes in a genome, and then (or simultaneously) to plot the sequence of bases within each gene, and the sequence of bases in the noncoding regions within and between each gene. Several organisms have now been sequenced in this way, and a few months before this book went to press the Sanger Institute in Cambridge produced the first rough draft of the entire sequence of bases in the human genome. All this derives from the ability to cut up DNA into pieces by means of restriction enzymes.

But genetic engineering is not primarily a matter of analyzing the sequence of bases in DNA. It's about joining different bits of DNA together to make *recombinant DNA:* lengths of DNA that contain material from two different sources, which may be two totally unrelated organisms. So how are different bits of DNA joined together?

Well, some restriction enzymes simply cut the DNA straight across, leaving "blunt ends." But some make a staggered cut, typically leaving two bases poking out from each of the two cut ends. These spare bases are known technically and concisely as sticky ends. They are sticky because they have loose chemical bonds and so are highly prone to attaching themselves to other samples of DNA that happen to be around, provided they too have sticky ends. If two samples of DNA from different sources are mixed and exposed to the same restriction enzyme to create sticky ends, then they will immediately join up again with other sticky ends; and each kind of DNA is just as likely to rejoin with a bit of DNA of the different kind as it is with its own kind. The rejoining is mere chemistry, and chemistry is blind, whatever its biological consequences. This initial, passive rejoining is weak, however. The join must then be consolidated—"annealed" is the technical term—using another enzyme known as a *ligase,* which is normally concerned with DNA repair. Thus the genetic engineer employs the agents and the mechanisms that nature has already provided. The first recombinant DNA to be created in such an orderly fashion was made by Janet Mertz and Ron Davis in California in 1972.

DNA and enzymes (which of course are proteins) are also in dialogue. DNA provides the code that shapes the proteins, to be sure; but some of those proteins are then required to form the DNA. So which came first? This is a chicken-and-egg problem. Neither came first, is the only sensible answer. The primitive precursors of proteins and DNA surely arose separately, neither at first requiring the other. Then the two coevolved into their present, wonderfully sophisticated modern forms. Thus life is not innately hierarchical. It is inveterately dialectic.

By itself, the creation of recombinant DNA is not genetic engineering. Genetic engineering implies that a piece of DNA— in fact, a functional gene—is stitched into a chromosome within a new host organism. Conceptually, the simplest way to do this is to inject the novel DNA into the nucleus of the recipient cell, and this, until the mid-1990s, was almost the only method of getting new genes into animals. "Engineers" of plants, however, generally employ vectors, which carry the novel DNA into the new host and supervise its integration into the chromosomes. Viruses are often employed as vectors; otherwise, the most common vector employed in plants is the bacterium *Agrobacterium tumefaciens,* which in nature is a pathogen that causes crown gall. Bacteria typically practice a form of sexual exchange in which they pass packets of DNA known as plasmids to each other. *A. tumefaciens* normally passes plasmids into plant cells to induce them to form the galls on which the bacterium then feeds. Genetic engineers modify the *A. tumefaciens* plasmids to act as vectors. Herbert Boyer and Stanley Cohen in California first introduced novel DNA successfully into new hosts in 1973. Thus recombinant DNA technology was born in 1972, through Mertz and Davis, and 1973 saw the birth of "transgenesis," the transfer of genes between organisms, by Boyer and Cohen. These are the basic techniques of what is colloquially called genetic engineering.

Now, nearly thirty years on, genetic engineering has still not transformed the world's economies in the ways that have sometimes been predicted. Many of the predictions have been unrealistic: greatly exaggerated not by the much-maligned media, but primarily by the scientists and biotech companies themselves as they bid for grants and venture capital. Technologies of this power and significance take far more than 30 years to unfold. In 300 years our descendants will still be exploring the possibilities. But although these are early days in the history of genetic engineering, a great deal has happened nonetheless.

WHERE WE'RE AT: THE TRINITY
OF BIOTECHNOLOGIES

The ability to isolate genes, and to stitch them into new hosts, raced ahead in the seventies, eighties, and nineties. From the beginning, however, the engineers felt frustrated. Only in a few, rare cases did they know which genes—which particular stretches of DNA—they should be transferring into the organisms they wanted to transform. They might know that one particular potato probably contained a gene that conferred resistance to, say, mildew because that particular potato was indeed resistant. But that was a fact of classical genetics, the kind of genetics that treats genes as abstractions. As engineers, they needed to know which particular piece of DNA corresponded to the resistance gene.

The attempt to find out which bits of DNA in an organism correspond to which genes is called *genomics*. It has three goals: first, to map the positions of all the genes on their various chromosomes; second, to work out the sequences of bases within the functional genes, and within all the bits in between; and third, to show what each gene actually *does*, whether making proteins, transfer RNA, or ribosomal RNA. Genomics has been the grand pursuit of the 1990s and will continue into the twenty-first century and beyond. The genomes of a few simple organisms have already been mapped and sequenced. The human genome is well on the way. The principal farm livestock will follow soon, and so will the world's major pathogens. In time, as the methods of sequencing become ever more automated, rapid, and routine, biologists will surely sequence genomes from representative species across the entire tree of life. These are exciting times. Among other things, this work will give us many surprising insights into the evolutionary relationships—the phylogeny—of our fellow creatures.

However, some claims made for genomics are a little overstated. It is not true, for example, that the printout of the human genome will enable biologists to "read the book of life" in the sense that we can now read, say, *Hamlet*. If we compare genes to words, then we should compare the genome as a whole to literature. The individual words interact; the meaning of the whole lies in the interactions; and the genome—like literature—is full of historical allusions, puns, nuances, caprices, abbreviations, colloquialisms, and so forth. Like any book that comes straight from the author, it has many a typo as well. The printout of the genome will provide the equivalent of a dictionary, a lexicon. But to "read" it we need to know its syntax and all its quirks of dialect. We could not understand *Hamlet* if all we had was a dictionary. It will be hundreds of years before we can interpret the human genome in the way that we can now interpret *Hamlet*. Some already talk of "designer babies," implying that the human genome might be manipulated to produce people to prescription. Would anyone undertake to edit *Hamlet* if all they had was a dictionary that sketched the rough meaning of individual words? Would anyone—to make the analogy more cogent—undertake to edit an epic poem written in Linear B? Not if they were sane, is the short answer.

Genomics, nonetheless, is the necessary handmaiden of genetic engineering. Without it, the technology of recombinant DNA remains a laboratory exercise.

So far, too, genetic engineering has advanced most rapidly in bacteria, producing, for example, novel antibiotics, and in plants, producing a wide range of new crops of the kind known as GMOs, or genetically modified organisms. There have been only a few successful manipulations of animals, and very few of direct commercial significance. One of the first that truly has been successful is "Tracy," a sheep that was fitted with a human gene that produces the enzyme alpha-1 antitrypsin, or AAT, which is used to treat cystic fibrosis and emphysema. Tracy produced AAT in her milk, and her descendants are now producing AAT in amounts that should soon prove to be of huge medical significance. That at least is the realistic hope of the biotech company PPL, based near Edinburgh, which owns and is developing Tracy's descendants.

The transgenesis of animals has lagged behind that of bacteria and plants for various reasons. One is welfare. Until recently, novel genes had to be introduced randomly into the genomes of the host animals, and such importunate introductions can cause changes in other genes, with untoward effects. Then again, some genes have highly circumscribed effects, inducing resistance, for example, against a particular pathogen. But others—those that affect overall growth rate, for example—influence the whole metabolism and so may throw the whole organism out of kilter. Most genes are pleiotropic to some extent, and so again can produce unexpected side effects. For all these reasons, genetic introductions can produce very distressed-looking organisms. If the recipient is a plant, then the distress does not matter. Most of us assume that plants do not have feelings. But if an animal is distressed—a monster, a late abortion, neonatal death—then we should take pause. This is another reason that the genetic engineering of humans looks less promising than some enthusiasts seem to suppose.

The second reason is more technical. In some ways, it seems easier to get novel genes into animal cells than into plant cells. Plant cells, after all, have thick walls of cellulose, and most of the insides tend to be water, with the nucleus tucked away to one side. Animal cells lack walls and are much more compact, with the nucleus occupying proportionately more of the space. This is why it is possible simply to inject novel DNA into animal cells, while plant cells require vectors.

But plants do have one huge advantage. Ever since the 1960s it has been possible to multiply plant cells in culture, as if they were plates of bacteria, and then regenerate entire new plants from single, cultured cells. This is not the case with all species but it is becoming feasible with more and more. Thus it is possible to introduce DNA into hundreds or thousands of cells at a time, as they lie in culture. Then, by attaching suitable markers, the biologists can pick out the cells that contain and express the introduced gene and make new plants from them. They can easily make thousands of attempts to produce just one good, functional, transgenic plant.

Until recently, the only way to make transgenic sheep, such as Tracy, was to inject DNA directly into a one-celled embryo in vitro, then implant that embryo into the womb of a surrogate mother. Sometimes this worked, but usually it did not. It took an awful lot of embryos to produce one in which the introduced gene was expressed as required. The technique was expensive and time-consuming, and hardly practical. Tracy was a rare success.

This has now been changed by the techniques of cloning developed at Roslin Institute near Edinburgh, which is a government laboratory, together with its neighboring biotech company PPL. First, in 1995, Keith Campbell and Ian Wilmut at Roslin produced two sheep, Megan and Morag, that they grew from embryo cells they had multiplied in culture. The following year, 1996, they produced the famous Dolly, who was grown from an *adult* cell (taken from the mammary gland of an old ewe) that had been multiplied in culture. Such cloning alone is a tremendous achievement. All animal cells differentiate in culture, and the mammary gland cell from which Dolly was cloned was, of course, highly differentiated to begin with. That whole adult animals were produced from these cultured cells demonstrated beyond doubt that the genomes of differentiated animal cells *can* be reprogrammed, provided the conditions are appropriate.

Most commentators at first assumed that cloning was the *point* of the Roslin and PPL experiments. Not so. The real point was and is that once the cells are in culture, genetic engineers can work on them just as they already can with cultured plant cells and bacterial cells. They can add DNA to hundreds or thousands of cells not by injecting it, but simply by pouring it over the top (though there are refinements). Then the engineers can pick out the cells that have taken up the required gene and are expressing it. They, too, can now make thousands of attempts for every transgenic animal they produce, a far cry from the days when they had to inject one-celled embryos and had only one try per animal.

In 1997 this technique was taken to completion with the birth of the ewe Polly. She was produced from fetal cells that were transformed genetically in culture and is the first mammal to be both cloned *and* genetically engineered. She carries the gene that produces human blood-clotting factor IX, which she secretes in her milk (just as Tracy's descendants secrete AAT). Factor IX is used to treat the form of human hemophilia known as Christmas disease. At present factor IX is obtained from human blood, with all the attendant disadvantages of expense and possible transmission of infection. Production of therapeutically useful protein in the milk (and potentially in other tissues) of livestock has been called pharming. Pharming, at present, is one of the most promising uses for the techniques of cloning and genetic engineering.

Or, to put the matter the other way around, cloning by cell culture and nuclear transfer is the technique that was required to make the genetic engineering of animals a feasible technology. Biotechnologists will soon be able to produce transgenic animals as readily as they now turn out transformed plants and microbes.

Thus, genetic engineering is coming of age by virtue of two subsidiary biotechnologies: genomics and cloning. These three—genetic engineering, genomics, and cloning—are the modern trinity.

Of course, cloning is an intriguing technology in its own right, even without genetic engineering. Many have envisaged that humans might be cloned, although this may well remain a fantasy, as it probably should. American researchers in particular have sought to clone "elite" cattle, to raise the general standard rapidly.

I believe, however, that the most significant role of cloning may be in animal conservation. Even in the best-managed programs, herds of endangered animals will have lost at least some of their genetic diversity in 100 years' time. Suppose, though, biologists now rushed around the world as fast as possible, gathering tissue samples (just a few cells will do) from well-chosen samples of all the animals they can lay their hands on, both from zoos and from the wild. These tissue samples can then be cultured and then deep-frozen in liquid nitrogen. (They have to be cultured *before* they are frozen to produce cultures that are just one cell thick and can be frozen evenly and quickly, without damage.) In 100 years' time, when cloning culture is more efficient, the remaining members of the species—or even some individuals from some closely related species—can be given embryos reconstructed using the cells that have been frozen now. With such technology, it will not be necessary to pursue the elaborate protocols that are now required to reduce genetic drift. Virtually all the alleles that now exist within any species could be kept permanently in store; all that will be needed to turn them into whole animals are healthy females of the species in question. Conservation zoologists—including all zoos—*ought* to be collecting tissue samples now, as a matter of urgency. To my knowledge, however, few are doing so.

All in all, the power of genetic engineers, abetted by the subsidiary techniques, is becoming stupendous. As the decades and centuries pass, they will be able to create novel creatures to order, and at will. Such power is not necessarily bad, but it must clearly not be taken lightly.

Now we will see how all the technologies discussed in this chapter have been essayed in human beings—or might be in the future, if some people have their way.

NINE

The Shaping of
Homo sapiens

Could we breed "better" human beings, just as we have "improved" our crops and domestic animals? Could we make people who are more handsome, more resistant to disease, more intelligent? Could we, as we develop the science and skills of genetic engineering, finally produce the "designer baby," specified to the last unit of IQ, the finest tilt of the nose? Many people in recent history—including a significant army of intellectuals—have found such questions enticing. In the light of twentieth-century events, they are more likely to seem chilling. But they must be asked.

The designer baby *must* be considered technically possible. We could not at present make new human beings to specification, but as things are, there seem to be no absolute barriers or none at least that are posed by science. Over the past four decades the idea that there *are* absolute barriers posed specifically by biology has become outmoded. In the 1960s most scientists still assumed it was impossible to transfer genes between unrelated species, yet by the early 1970s we entered the age of genetic engineering. In the mid-1980s a world-renowned embryologist declared in *Nature* that it was "biologically impossible" to clone a mammal by nuclear transfer, but a decade later scientists at Roslin Institute produced

Megan and Morag, and then Dolly, in exactly that manner. The transfer of genes and the cloning of mammals from somatic cells: no biological barriers that have ever been envisaged seem more profound than these, and yet they have been overcome. Nowadays it is most prudent and most honest simply to presume that *anything* in the realm of biotechnology should be considered possible provided we do not attempt to break what Sir Peter Medawar called the bedrock laws of physics, or to defy the rules of logic. There is nothing physically or logically impossible about designer babies. We should start asking as a matter of some urgency what might really be implied by the idea. In the next chapter I will look at the ethical issues. This chapter is about the practicalities.

Even without officious intervention, of course, our gene pool will not stand still. DNA is innately restless. Clearly, too, our own lineage, leading to *Homo sapiens,* has evolved rapidly over the past few million years. So how might our species change in the future if we eschewed all talk of "improvement" and simply let nature do as nature does?

HOW MIGHT HUMAN BEINGS EVOLVE IF WE LET NATURE TAKE ITS COURSE?

That human beings did indeed evolve—and that we share a common ancestor with the apes—was one of Charles Darwin's most shocking revelations, and now it seems that our lineage separated from that of the chimpanzees only around 5 million years ago. Our oldest known specifically "hominid"—humanlike—ancestor was *Ardipithecus;* then came *Australopithecus;* and from those australopithecines, just over 2 million years ago, emerged the first of our own genus, *Homo.* The genealogical tree of *Homo* has branched, like all evolutionary trees, but the particular line that led to us seems to run from *Homo habilis* to *Homo ergaster* (who was very like the better-known *Homo erectus*), who in turn gave rise both to us,

Homo sapiens, and to our "sister species," the Neanderthals, *Homo neanderthalensis.* The various species of *Australopithecus* walked upright but were short—generally not much more than a meter—and had small brains, around 450 milliliters. *H. habilis* was a little taller (not much) but had a significantly bigger brain, around 700 ml. *H. ergaster* was as tall as we are, with a brain of around 1,100 ml. Our own brains average around 1,400 ml, while those of the Neanderthals were bigger still, but then, the Neanderthals had bigger bodies as well.

It has become unfashionable and politically incorrect to suggest that there is anything resembling progress in evolution (although there clearly is, once we distinguish "progress" from "destiny," but that is another story). Yet the *trend* is undeniable. Since the human line departed from the chimp line, 5 million years ago or so, our own ancestors became more upright, taller, and distinctly brainier. Our brains are three times as large as those of chimps, or indeed those of the early australopithecines.

Folklore commonly has it that those trends—particularly the increase in brain size—will simply continue. The favorite boys' comic in 1950s Britain was *The Eagle,* featuring spaceman Dan Dare, "the pilot of the future," whose archenemy was the Mekon: a green homunculus from Venus with a head the size of a dustbin. Are we all on course to become Mekons? Will our brains continue to grow, as they have done in the past? Will our legs shrink, as we travel more and more by car?

The short, quick answer is no. There are many reasons for supposing that in the past, survival, sexual, and social selection combined to drive our evolution from *Australopithecus* to *H. sapiens.* The appropriate pressures were there: a big-brained animal with hands to translate its thoughts into action would surely have outcompeted its slower-witted contemporaries, attracted more mates with its tricks and jokes, and found it easier to socialize with others who were similarly blessed. In addition, it is easy to see how those with slower wits (reflected in smaller brains) simply died out. In a state of nature the pressures are enormous; anyone who could not withstand those pressures was pushed aside. Nowadays, extra intelligence still brings clear advantages, including a well-paid job as, say, a New York lawyer. But the people who cannot outsmart the New York lawyers do not fade into oblivion as they might have done when we all competed on the African plain. They are more likely to employ those lawyers to fight their battles. To be sure, many people are still under enormous pressure and die before their time from disease and poverty. But—genetically speaking—such blows strike more or less randomly. A very intelligent and athletic person living in tropical Africa is probably more liable to die young than a much less alert individual who happens to live in the English home counties.

This does not mean that natural selection has ceased to operate. It is a fact of the universe and cannot be escaped. Natural selection is ensuring, for example, that certain forms of disease resistance are increasing among some sections of the human population. But the consistent driving pressure that turned *Australopithecus* into *Homo ergaster* and then into *H. sapiens* is no longer bearing down on us. Society still favors intelligence, but overall, the big-brained people are not outbreeding those with smaller brains. There is no consistent pressure, at present, to make us more Mekon-like.

However, the Mekon does provide a nice demonstration of biological impossibility, because he really would break various laws of physics. His enormous head was clearly too heavy for his spindly neck—defying the laws of mechanics—but his worst problems were thermodynamic. Brains are incredibly greedy organs: it takes about 20 percent of our total energy to service our enormous nervous systems. If the Mekon's brain was ten times as big as a modern human's, it would consume at least twice as much energy as an entire person. He would need an enormous gut to feed such a monstrous organ, yet he was a physical weed. Such a brain would also generate enormous heat and would be very difficult to cool.

His huge skull would need to include a refrigeration unit. On a quite different tack, our personal failure to exercise would not in itself cause our legs to shrink in any heritable way. August Weismann pointed out that changes wrought in our bodies are not reflected so simply in our gametes. More broadly, such an evolutionary progression would be Lamarckian, and Lamarckism does not work—or not, at least, in this gross fashion. In practice, we are not growing smaller, but a great deal taller. As discussed later, the reasons for this may *seem* Lamarckian, but they are not.

In fact, there are various reasons to suppose that *Homo sapiens* is likely to stay much the same for the next few thousand years—or indeed for the next few million—unless we set out officiously to change it by breeding or engineering. The fossil record shows that many creatures go through long periods when they hardly change at all; some lineages of clams remained virtually the same for tens of millions of years. Change, when it does come, is often rapid, giving the pattern that Stephen Jay Gould and Niles Eldredge have called punctuated equilibrium. Our own fossil record suggests that we are in a period of stability; certainly, the fossils indicate that our ancestors have hardly changed at all in the past 100,000 years, at least anatomically, although they clearly evolved very rapidly at times in the few million years before that.

Evolutionary change implies alteration of the gene pool: the total inventory of alleles contained in all the genomes of the population. The total DNA in each human genome amounts to around 3 billion base pairs, which in turn account for around 34,000 genes. Yet the range of allelic variation within the world's current population of 6 billion people is remarkably small, less than is known to exist even within the remaining few thousand chimpanzees. The reason, so some suggest, is that the human species went through a severe genetic bottleneck around 70,000 to 80,000 years ago, when the total world population must have been reduced to a few thousand. (One possible cause of this putative reduction was the eruption of Toba, the "supervolcano" on

Sumatra, which is known to have occurred around 74,000 years ago. Such huge volcanoes send so much dust into the stratosphere that sunlight is all but excluded for several years, bringing on a severe if short-lived ice age and killing just about everything. However, this scenario does not explain why other creatures, who on the face of things were no less vulnerable, apparently did not suffer a similar fate.)

There are good reasons to suppose that our present gene pool is as stable as gene pools are liable to get. Of course, the frequency of some alleles is bound to be changing all the time, and some would say that such change is evolutionary by definition. However, we should surely distinguish between the kinds of changes that may be temporary, and might in principle be reversed, and those that cause permanent, significant change, perhaps leading eventually to speciation. The first kind of change is bound to happen constantly. Kenyans at the moment are breeding faster than Italians, for example. The Kenyan population is currently on course to double every 25 years or so, while the Italian population is, if anything, going down. So the proportion of alleles typical of Kenyans is increasing relative to the proportion of alleles typical of Italians. Of course, *most* of the alleles possessed by Kenyans are the same as those in Italians, so fluctuations in the two populations make no overall difference. But some of the alleles in the two groups are different. (If they were all the same, the two groups would look much more similar than they do.) Yet it seems silly to class such fluctuations as "evolution." They could, after all, prove reversible. One day the Kenyan population might stabilize, while the Italians could again become pronatalist. We surely should not speak of "evolution" unless there is a clear, qualitative shift: a loss or gain of alleles that is not reversible.

Beyond doubt, some irreversible qualitative changes in the human gene pool must be taking place as well, but these on balance seem too small to be significant. On the one hand, the human gene pool must now be expanding. New mutations are occurring the whole time, and with 6 billion people now on board—and the population still rising—the increase in mutations must be faster than ever. Some of those mutations are occurring in the germ cells and thus will be heritable. We are also losing variation as aboriginal tribes, who are likely to contain rare alleles, dwindle and disappear. But such gains and losses, however sad the losses may be, strike randomly (genetically speaking) and on the whole are marginal. All in all, then, if we do not contrive deliberately to alter ourselves, we seem likely to stay much as we are.

Yet there is no fundamental principle that says our evolution is *bound* to come to a stop, even though such a notion remained fashionable among some biologists well into the twentieth century. For although those biologists paid lip service to Darwin's idea that human beings and apes had evolved from a common ancestor, they continued to believe in their hearts that we had after all evolved "in the image of God," just as it says in Genesis. Many simply could not bear to believe otherwise. Now that we had attained godlike stature, we surely would evolve no further. Robert Broom, the distinguished paleoanthropologist who discovered *Australopithecus robustus* in southern Africa in the 1930s, thought precisely along these lines. He had found direct evidence of human evolution in the past, but, he said, now that *Homo sapiens* had come into existence, we should expect no further change in the future because God's purpose was already fulfilled.

Such a view is sublimely nonbiological. There is progress in evolution, but there is no destiny, and our present condition cannot be taken as our denouement. If the condition of the world changes radically—if there is ecological disaster—then quite new kinds of hominids could arise from among our ranks. Global warming, all-out nuclear war, or some combination of the two could do the trick. We know that volcanoes can be very effective. There is at present a vast and smoldering supervolcano beneath Yellowstone National Park in Wyoming that is apparently due to erupt at any time. Disaster on such a scale—perhaps exacerbated by a rise in fundamentalist religion, with renewed antipathy to science—could destroy modern communications and break the world's population into fragments, just as it was fragmented in Pleistocene times. Each fragment would then be subject to different ecological pressures. If the isolated groups contained sufficient genetic variation, and if the pressures were consistent but not so strong that they caused extinction, then each group could readapt in whatever way was required to its own novel conditions.

But there is no outstanding reason to suppose that these newly isolated groups would continue to evolve along the lines of the past few million years. Future selective pressures might favor *smaller* brains that need less energy. Our scattered descendants might also find it convenient to climb into the trees to avoid tomorrow's predators, who surely would be enjoying a field day. Our post-disaster descendants would not turn back into the kinds of apes that are already familiar to us, but in a general way, they could well become more simian again rather than less. Then again, they might go off in new directions altogether. Our present form was not inevitable, and it cannot be seen to be permanent. We should enjoy ourselves as we are, while we may.

Even so, in the absence of global disaster we do seem most likely to stay as we are, unless we set out deliberately to "improve" ourselves. But how might we do such a thing?

THE PATHS TO IMPROVEMENT

Among the paths to the biologic "improvement" of human beings we can distinguish two kinds of trend. First, we can envisage a steady increase in ambition. Most basic are the traditional medical concerns of therapy and prevention: the simple desire to alleviate or prevent the suffering caused by deleterious alleles. But after that, we may envisage a steady crescendo of endeavors not simply to ameliorate frank disorder, but to improve ever so slightly, notch by notch, on human beings who by reasonable standards are perfectly healthy already. Finally, we might in principle design a human being absolutely to specification, in the way that Ferrari now designs cars. This final analogy is not quite accurate, of course, because the designer of living creatures, human or otherwise, must be more like a chef or a gardener than an engineer. In living systems more than in machines, each individual part interacts absolutely with every other; the organism must be conceived as a whole. Indeed, although chemists use the term *organic* simply to mean "carbon-based," in common parlance it refers to systems in which each part depends absolutely on all the others and the whole is greater than the sum of the parts. In living systems, too, as in gardens and great recipes, the final criteria of excellence are aesthetic. Of course, designers of fine cars would feel insulted by these musings; they would argue, quite rightly, that it is also necessary to conceive of great machines as a whole and to be guided by aesthetic principles. Even so, it is a matter of degree. Living systems are more complex than human artifacts, and are more "organic" in the common sense. Nonetheless, the engineering analogy remains. In principle, future biologists might build human beings to the same precise specifications that are now employed to shape cars.

The second trend is in the steadily increasing power of the technology. The oldest attempts to change human beings (or to keep them as they are) predate Mendelian genetics and simply employed commonsense notions as acknowledged by traditional breeders. The twentieth century saw increasing application of both Mendelian genetics and—at least as conspicuously—the parallel tradition of Francis Galton's biometrics. In the second half of the twentieth century, and increasingly toward the end, we saw the steady growth of reproductive technologies that, though not directly intended to alter the human gene pool, are vital for any future, high-tech interventions. Artificial insemination came first: an ancient technique, but made practical in the 1950s when it became possible to freeze sperm (so that donor and recipient did not have to be in adjacent rooms). Then came in vitro fertilization, or IVF, in the late 1970s: conception in a petri dish (though more usually represented journalistically as a test tube). Since then we have seen a flurry of reproductive techniques, culminating in cloning. Although only sheep and a few other species have been cloned so far, no one doubts that this technology is applicable to people. Cloning, as discussed earlier, makes it feasible to carry out precise and large-scale genetic engineering in animals, including humans. The animal cells can be manipulated in culture, just as is now possible with plants and microbes. The highest of high technologies are already—in principle—in place.

The two kinds of trend—in ambition and in technique—are entirely independent of each other. The high technologies are already being used in orthodox medicine in the effort simply to reduce suffering. On the other hand, the most basic techniques of pre-Mendelian breeding could in principle be used to make dramatic transformations—that is, to breed "better" human beings. This, essentially, is what the Nazis attempted as they contrived to create a new, more emphatically "Aryan" human species.

Clearly, it would be possible to write a very large book on the hierarchy of possible ambitions and technologies; but I will pick out just a few salient items.

The most obvious target for prevention and therapy is the 6,000 or so known deleterious alleles that can cause single-gene disorders. Most of these are recessive, and most of the recessive ones—the ones that are not specifically linked to X chromosomes—cause disease only when inherited in a double dose. The easiest way to prevent most of them, therefore, is for carriers—heterozygotes—to avoid having children together. This notion forms the basis for counseling, or at least for much of it.

In fact, counseling can be carried out at a significant level without any science at all. Common sense can do a lot. Any pre-Mendelian nineteenth-century breeder of sheep could have told the royal houses of Europe not to marry among themselves, since some of their members suffered what seemed to be heritable diseases, notably porphyria and hemophilia. But who listens to breeders of sheep? In the twentieth century, knowledge of Mendelian ratios and access to family trees enabled generations of professional counselors to be a great deal more precise; in some cases they could at least work out the odds that particular people might be carriers of particular alleles, and the odds that their children might be affected.

We are only talking odds, however. The whole exercise becomes more satisfactory when it becomes possible to detect carriers directly, either by biochemical means (small changes in the phenotype) or by looking directly for genetic "markers" in their DNA. The search for carriers is in turn reinforced by detection of the homozygous state in the embryo itself, at a stage when it is still possible to do something about it. Diagnosis can be made at birth, which is how pediatricians have been diagnosing alkaptonuria, the disease that enabled Archibald Garrod to define the function of the gene. This is worthwhile because alkaptonuria, if detected early, can be treated effectively. After the mid–twentieth century it became possible to detect at least some disorders in utero, by looking for loose cells with the technique known as amniocentesis. Now it is possible to detect a wide range of potentially damaging alleles in embryos conceived in vitro, allowing the option of implanting only those embryos that are free of the deleterious allele.

This last approach is very encouraging: to me, it truly looks like high tech in the service of sensible therapy. It is now becoming possible, after all, for two known carriers of, say, cystic fibrosis to produce offspring in the *absolute certainty* that their child will not only be free of the disease but will not even be a carrier. Embryos are produced by superovulation and in vitro fertilization, and the ones that do not contain the deleterious allele are selected for implantation. Some people feel that such an approach is too interventionist, or indeed callous. However, nature herself is extraordinarily profligate with human conceptuses (although, intriguingly, much less prodigal with those of animals that breed seasonally, such as red deer). Most human embryos abort within the first few weeks. Abortion in a petri dish, before pregnancy has even begun, does nothing that nature does not do as a matter of course. More generally, although intellectuals of all kinds—including scientists and moral philosophers—feel it is their job to improve on common sense, I feel that we override common sense at our peril. Common sense says that—given a choice—it is better to give the chance of life to an embryo that has the potential to be perfectly healthy than to one who will be seriously ill from day one. In vitro fertilization, followed by embryo selection, seems to me to be a very good approach to serious single-gene disorders, provided of course that there is no coercion and that everything is done in accord with the parents' wishes. (The embryo ought to have a say too, of course, in a perfect world. But we have to work within the limitations that the world imposes.)

One flaw with all these techniques, from pre-Mendelian counseling to in vitro diagnosis, is that they tend to be post hoc. That is, nobody would counsel potential carriers of cystic fibrosis, say— and still less would they attempt in vitro diagnosis—unless they had good reason to suppose that the people concerned *were* carriers. They would already have a family record of disorder and perhaps already have affected children. It would surely be better to intervene *before* there is a family record of disease. This, of course, leads us into "well-population screening": looking for disorders in people who seem perfectly healthy. Already this is possible for some disorders: where there is a straightforward test; where there is good reason to suppose that people *might* be carriers (even though there may be no direct evidence); and where the resulting disease is serious. Thus, many health authorities offer mass screening for sickle-cell anemia within people of African descent. In America, Ashkenazi Jews offer screening for the allele that leads to Tay-Sachs disease. As Matt Ridley points out in his excellent book *Genome* (whence I garnered this example), there is no coercion in this. People themselves decide what their problems are and decide appropriate solutions, and no individual is obliged or in any way pressured to take up the available technologies.

In time, of course, it will in principle be possible to screen *everybody,* for all deleterious alleles, just by speed-reading their DNA. The Sanger Centre in Cambridge has already provided the first draft of the entire human genome, showing the sequence of nucleotides in most of it (all but the last, trickiest sections). Over the next few decades it will become possible truly to "read" that sequence and to see precisely which individuals contain which alleles. In addition, the techniques for sequencing have improved by leaps and bounds even since the Sanger first began its work just a few years ago; these are techniques that have been developed largely because of the Sanger's endeavor. Within less than a century (probably far less), local clinics will be able to offer a readout of anyone's DNA within hours. Doubtless, indeed, this will form part of everyone's standard medical records. Then, people who wish to have children together can compare their genomic readouts to see if they share any alleles that could cause disease if combined. Some may find this chilling. The main point, though, is that the decision to match a couple's genomes, and on what to do subsequently, should remain with the couple. It becomes sinister only when it becomes the business of the state.

For a child born with a genetic disability, it is becoming possible to correct the damage. Therapists envisage, for example, that defective tissues might be removed, the deleterious alleles replaced, and then the tissue restored. Such treatment has been envisaged for cystic fibrosis for at least two decades and will become feasible sooner or later.

This approach, however—correcting damaged genes and hence the tissues in which they are expressed—raises the possibility of a huge conceptual leap. For if a doctor simply corrects the genes (and hence the cells) in the lungs of a person with cystic fibrosis, then only the patient is directly affected. The genetic changes made in the lungs are *not* transmitted into the gametes for the reason August Weismann identified more than a century ago. Thus the children of the treated person are not affected.

It would also be possible in principle, however, to correct genetic defects in embryos in vitro (provided of course that they were produced by IVF). If this were done, then the genetic changes made could well find their way into the developing germ tissue, and hence into the gametes, and hence would affect all subsequent generations. This is what is known as germline therapy.

Some commentators feel that there is simply no sensible reason to undertake germline therapy, at least in this form. After all, when embryos are conceived by IVF they are not made one at a time. Typically, the woman produces about a dozen eggs after superovulation, resulting in about a dozen embryos. It is possible to pick out the embryos that do not contain a double dose of the damaging allele, or that, indeed, contain no damaging allele at all. It is certainly easier to do this than to correct genetic damage in the embryo. Thus to me (and many others) it seems perverse even to contemplate this kind of germline therapy. Why attempt heroic correction of a diseased embryo when it would be much easier to offer the chance of life to a healthy one? However, germline therapy of this kind does have its advocates, who propose, for example, to correct cerebral palsy and to eliminate it from all of the sufferer's descendants. Germline therapy is not yet technically possible, though it is in principle feasible. So we do have time to think about it.

However, all these attempts to preempt or correct single-gene disorders will *not* appreciably affect the frequency of those alleles in the gene pool. Counseling, with or without high-tech examination of DNA, prevents carriers from having children together, or at least gives them the option. But it does not eliminate the alleles they are carrying. By the same token, treating patients with cystic fibrosis so that they live to be adults and have children of their own does not appreciably increase the frequency of that allele. For one thing, as we have discussed earlier, the number of heterozygotes—carriers—greatly outweighs the proportion of homozygotes. For another thing, by the use of in vitro or "preimplantation" diagnosis, it is possible for sufferers to produce offspring who themselves are free of the diseases. (If the sufferer mated with someone who was not a carrier, then all the children would be carriers but would not be affected. If the sufferer mated with a carrier, then half the children would be homozygotes, but could be identified in vitro. The other half would be carriers. The offspring of those carriers could then be diagnosed in vitro, and, as we have already seen, some of the offspring even of two carriers should be completely free of the disorder.)

Thus the notion that prevention of single-gene disorders significantly "improves" the human gene pool as a whole is fatuous, at least within a reasonably measurable interval, and so is the idea that it will compromise the human race if we treat sufferers humanely and enable them to have children of their own.

This brings us to another great conceptual leap: from attempts to correct or prevent obvious suffering to attempts deliberately to improve on creatures—human beings—that are perfectly healthy already. This takes us into the realm of eugenics.

EUGENICS

Different people seem to mean different things by "eugenics," but I perceive it as the attempt to improve human beings by adjusting the human gene pool as a whole, as well as the genomes of individuals. People will also differ, of course, in what they mean by "improvement." The general approach, however, must be to reduce the frequency of alleles that are seen to detract from the intended goal, and to increase the frequency of alleles that seem to lead in the required direction. As with conventional medicine, the methods may be low tech, simply a matter of arranging some matings and discouraging others. Or they might in future involve the highest of high technologies, including all the techniques of genetic engineering, including the addition of novel genes synthesized in the laboratory. Thus in principle the idea of eugenics embraces the notion of the designer baby, even if that is not what its founders or its present protagonists necessarily intend.

Some of the practices of conventional medicine may *seem* to be leading us in eugenic directions. Some critics of gene therapy, prenatal diagnosis, and even of genetic counseling suggest this. Yet, as we have discussed, gene therapy as currently envisaged is merely intended to reduce the suffering of individuals; even if extended into germline therapy, it would still merely reduce the suffering in a particular family. Prenatal diagnosis and genetic counseling are intended primarily to reduce the chances of producing babies that are liable to suffer from single-gene disorders. These techniques do not significantly reduce the frequency of those alleles in the gene pool as a whole, so they cannot properly be called eugenic. Human suffering is reduced, but the human species as a whole is unchanged.

It is often suggested, too, that we all practice "eugenics" up to a point because we all practice mate choice. We all have strong views about who our sexual partners should be. Mostly, of course, we are keen to ensure that the other parent of our children is somebody we actually like, since it is necessary to form a relationship with that person. But also, at least subconsciously, we have some concept of how our children will turn out, and any such preference (some argue) is a prescription of a kind, and so has eugenic connotations. In many societies, of course, the elders arrange the marriages of the young men and women, and many others who do not necessarily go that far nonetheless have strong views on who their children *ought* to marry. Orthodox Muslims emphatically prefer their young men and women to marry other Muslims. Orthodox Jews discourage marriage to gentiles. Such exclusivity seems to imply some preconception of what the nature of the next generation ought to be. That, again, has eugenic connotations.

Even so, mate choice and arranged marriage seem to differ clearly enough from eugenics, at least from eugenics in its strong form. For orthodox Muslims are primarily concerned to ensure that their own kind is perpetuated, believing this to be the will of Allah. They do not suppose that the next generation should be measurably *better* than the present one, and the one after that better still. But this is implied in eugenics: that the lineage should change over time, steadily moving toward some target that is *different from,* and measurably better than, what exists at present. To be sure, eugenicists may not aim to produce supermen and superwomen, in the way that cattle breeders dream of the bull that puts on 5 kilos of beef a day and the cow that produces 20,000 liters of milk a year. But the eugenicist does want to produce a greater proportion of people who conform to some preconceived ideal: more people with an IQ of 140; more who are tall and fair; and so on. Indeed, eugenicists in general seek to shift the entire human species toward some preenvisioned target. This is clearly conceptually different from the desire of most tribes and societies in the history of the world, which is simply to perpetuate their own kind.

Finally, decisions on genetic counseling or gene therapy *should* be made by individuals, on behalf of themselves or the children they intend to have. This is not always the case. As we will see, many governments in the twentieth century took such decisions into their own hands. But personal choice, nonetheless, is the name of the game. People also choose their own mates. Even when marriages are arranged, it is usually by parents, or elders who are perceived to have special rights.

We could envisage, too, that changes of a eugenic nature could be brought about by individuals, exercising their own choices on their own behalf. In principle, the free market could achieve this. Women could decide to produce babies that conform more and more closely to some preconceived ideal by buying the semen of Olympic athletes and Nobel Prize winners from elite sperm banks. Lee Silver of Princeton University envisages in *Remaking Eden* that future parents might pay genetic engineering clinics to manipulate the genomes of their offspring. These are at least possibilities and as such should be considered.

Historically, however, early attempts at eugenics in the twentieth century have definitely been coercive. Many a government has taken it upon itself to prevent particular marriages, and sometimes specifically to encourage others. We cannot in fairness suggest that eugenics *necessarily* implies coercion, or is necessarily the handmaiden of totalitarianism. There is still a perfectly respectable eugenics society in London, whose protagonists very properly point out that *all* technologies, and indeed all philosophies, may be misused, and most have been, at some time. We should not condemn eugenics out of hand, just because its ideals have been misapplied. Opponents, however, point out that some technologies and philosophies lend themselves more easily than others to misuse. Eugenics has lent itself to coercion, and sometimes has been the tool of despots, partly because it really does lend itself to such deployment. In the same way, although guns and bombs can be used as agents of peace, we should not be overly surprised when in practice they are used to make war.

It would be hasty to condemn eugenics out of hand, and grossly unfair to suggest that everyone who takes eugenics seriously has evil intent. Nonetheless, it seems a dangerous philosophy, a little too hot to handle; and once we move beyond the realms of counseling and therapy, it does seem hard to justify any concerted effort to manipulate the human genome, or the gene pool as a whole.

Still, though, we should look at the ideas and the history of eugenics as dispassionately as possible. They are important, and they will not go away.

The term *eugenics* was coined in 1885 by Francis Galton, polymath, statistician, and Charles Darwin's cousin. Thus the idea did *not* arise out of the science of genetics, because Mendel was the only bona fide geneticist before 1885, and at that time his work had been shuffled to the sidelines. Galton's eugenics had two more primitive origins. On the one hand, it was rooted in his own "biometrics": the attempt to measure and quantify the visible features of living creatures and their behavior, an attempt that was presaged by phrenology, the analysis of cranial bumps. On the other hand, Galton applied the ideas of traditional breeders. In effect, he reasoned that if you could create better cows by selective breeding, then you should also be able to breed better human beings. The prefix *eu-* means "good" or "well." Thus a eukaryote is a creature whose body cells contain nuclei (*karyon* being Greek for "kernel," or indeed for "nucleus"). *Eugenics* therefore implies "good birth" or "good breeding."

For the first few decades of the twentieth century eugenics was very fashionable indeed. It was new on two fronts: Galton had only recently framed the general idea of it; and as Mendel's ideas became more widely known, they *seemed* to root Galton's ideas in robust science. As Matt Ridley records in *Genome,* Galton's disciple Karl Pearson wrote to him in 1907 to say, "I hear respectable middle-class matrons saying, if children are weakly, 'Ah, but that was not a eugenic marriage!'" Eugenicists in the early twentieth century pursued both the options that had long been explored by breeders of livestock: to eliminate what they perceived to be the negative aspects of the human race, and to accentuate the positive.

But for the early eugenic enthusiasts, elimination of the negative did not, as now, focus exclusively on obviously deleterious mutants, like the allele that leads to Tay-Sachs disease. Instead, early eugenicists identified broad phenotypic qualities of the human race that they perceived to be undesirable. The leading target was "feeble-mindedness." Winston Churchill, when he was Britain's Home Secretary in 1911, expressed a widespread sentiment when he declared that " . . . the multiplication of the feeble minded [is] a very terrible danger to the race." Churchill had a way of expressing big ideas in fine style, even if, sometimes, they were bad ideas.

As Ridley describes it, *most* of the Protestant countries of the Western world adopted some kind of laws in the early decades of the twentieth century to enable government to sterilize those identified as feebleminded. Six states of the United States had passed such laws by 1911, nine more by 1917. Virginia retained such laws into the 1970s. In the twentieth century the Americans, arch-defenders of liberty, sterilized more than 100,000 of their citizens for the heinous offense of feeblemindedness. Canada, Norway, Finland, Estonia, and Iceland passed similar laws. The Swedes, with a much smaller population than America's, sterilized 60,000 of their sons and daughters. Most notoriously, of course, Germany sterilized 400,000, and at the outbreak of World War II, the Nazis murdered 70,000 inhabitants of mental institutions, who had already been sterilized, to make room for wounded soldiers. As all the world knows, too, the Nazis extended the criterion of undesirability beyond feeblemindedness, to include entire races. Genocide has been common in the history of the human species but has rarely been so systematic.

Britain *almost* passed eugenic laws, in 1911 and 1934. The first was vigorously opposed and beaten down by a heroic liberal member of Parliament, Josiah Wedgwood, a descendant of the great potter and another relative (by marriage) of Darwin. The second occasion was in 1934, but by then the Nazis were on the scene and enthusiasm was waning. In the early twentieth century, however, among the developed countries of northern Europe, only the Catholic states, or those with a strong Catholic influence, notably Holland, eschewed such laws altogether. Catholics, after all, see the body merely as the vehicle of the soul, which cannot be compromised by the body's infirmities and, besides, belongs to God.

The Nazis also sought to enact, as a matter of policy, the second of the two eugenic possibilities: to breed better human beings, which in that Teutonic context meant big, handsome, smart, and blond, by arranging marriages between suitable partners. As a matter of honesty, I should record that a German friend of mine knows a man—now of course approaching old age—who was born by such an arranged mating. It would be convenient to relate that he is a twisted ugly psychopath, but my informant tells me that in fact he is handsome, athletic, and charming, everything the Führer could have desired. Ah well.

It has been customary, in the late decades of the twentieth century, to decry the eugenic movement first as science gone mad, and second as a blatant, cruel, right-wing plot: an attempt by the ruling class to beat the lower orders into shape, by whatever means came to hand. But history does not bear this out. Many of the most fervent advocates of eugenics in its early days bore the flags of socialism, in its various hues. John Maynard Keynes, Beatrice and Sidney Webb, George Bernard Shaw, Harold Laski. H. G. Wells, generally perceived to belong to the left, wrote most chillingly and frankly that "the swarms of black, and brown, and dirty white, and yellow people . . . will have to go." Go meant "die," although, he added (doubtless with a kindly twinkle), "all such killing will be done with an opiate."

Matt Ridley, whose own politics are patrician—rooted firmly in the idea of personal freedom with personal responsibility—suggests that socialism lends itself particularly well to eugenics, since it so readily grants power to the state and places the perceived needs of society as a whole above the rights of individuals. As a matter of history, the USSR, which was the world's first large centralized but avowedly socialist economy, did not pass eugenic laws, though it did of course persecute minorities for essentially racist reasons, and, as Ridley points out, it did contrive to eliminate raftloads of intellectuals. As a matter of history, too, the most vociferous opponents of eugenics were middle-aged men with big mustaches and top hats, commonly perceived in our age, which is given to caricature, as old-fashioned fuddy-duddies: G. K. Chesterton, Arthur Conan Doyle, Hilaire Belloc. Chesterton wrote that "eugenicists had discovered how to combine hardening of the heart with softening of the head."

Scientists—including some of the greatest, and of left-wing persuasion—caught the eugenic fervor as well. Indeed, they did much to stoke the fires. R. A. Fisher, critic but admirer of Mendel, was a keen advocate. So was Sir Julian Huxley. So too was J.B.S. Haldane, who for some time was a member of the Communist Party. These biologists did turn against the idea, however, partly because they saw its misuse by the Nazis, and partly, too, for reasons of science. In particular, both the behavioral psychologists and the anthropologist Margaret Mead argued from their different perspectives that human beings were hugely influenced by their environment. Both lines of argument seemed to lend weight to John Locke's idea of the seventeenth century that the human mind is a tabula rasa. The research on which Margaret Mead based her ideas has since been questioned, and even ridiculed. She interviewed young girls in Samoa, but it seems her nervous and perhaps mischievous interviewees spun her some fine yarns. The behaviorists have often been criticized for insisting too strongly that animal (and human) psychology can be studied rigorously only by focusing on behavior—that we cannot directly observe thoughts and emotions, so we should not factor them into our accounts. This idea, among other things, undoubtedly compromised attempts to improve animal welfare. Many people seemed to conclude, after all, that science had effectively disproved the notion that animals think and feel, whereas the truth was simply that some scientists had decided for procedural reasons not to take thought and emotion into account, since they were not directly observable and measurable.

However, both Mead and the behaviorists helped dim enthusiasm for eugenics since they suggested that people really can change their ways and develop along vastly different lines, *whatever* their underlying genetic endowment might be. No modern geneticist would deny this. The notion that scientists who are interested in the genetic bases of human existence are "genetic determinists" is an absurd canard, at best lazy-minded. But in the early twentieth century nonetheless, emphasis on environmental influences served as an antidote to eugenic fervor.

Thus for political reasons (largely, but not exclusively, the example of the Nazis) and for reasons of science (essentially a shift of fashion), enthusiasm for eugenics waned. Several biologists in the late twentieth century wrote books attempting to jump on its grave, and castigating everyone who had ever been involved, or at least selected targets. The lesson was taken to be self-evident. Eugenics is bad. Anyone who so much as expressed an interest in it—anyone, that is, who did not spit blood at the mere utterance of its name—was held to be self-evidently reactionary. In some circles, this has become the politically correct attitude.

But it is the job of science to look dispassionately, to pick out the facts of the case. If the facts seem distasteful, it is meet to come to terms with them nonetheless, and unsafe simply to pretend that they do not exist. I do not want, in this brief account, to trot out all the political arguments for and against eugenics, which have already been rehearsed in extenso. But I would like to ask a few faux naive questions of the kind that are rarely raised at all and yet seem pertinent.

Why, first of all, were so many people carried away by the ideas of eugenics in the early twentieth century? Its advocates certainly were not stupid: they included leading intellectuals. They were not necessarily bad: Shaw and the Webbs, for example, are remembered in history as humanitarians, and although Churchill has lately been the victim of iconoclasm, he was certainly no monster. But although, in the early twenty-first century, it seems self-evidently foul to breed human beings like cattle, to our great-grandfathers' generation it did not. There are obvious reasons for the disillusionment: the sobering example of Nazism and the realization that human beings really are shaped by their environment as much as by their inherited biology. But why did our early-twentieth-century forebears grasp the eugenic nettle so enthusiastically in the first place?

I don't presume to know the answer, but I would like to offer two thoughts: one general and one specific. The general answer is that the nineteenth century saw a rising tide of enthusiasm both for science and for the technologies that arose from science. People were excited by the new chemistry and physics: James Clerk Maxwell, J. J. Thomson, Albert Michelson, who measured the speed of light, and Edward Morley. And of course they were excited by the new biology: the Germanic tradition of cell biology and embryology, and the philosophical revelations of Darwin. Some of the greatest architecture of the nineteenth century is that of the engineers: from Decimus Burton's palm house at Kew, and Joseph Paxton's Crystal Palace at London's Great Exhibition, to the iron- and then steel-framed skyscrapers of Chicago and New York of the 1880s and onward. The enthusiasts (albeit with reservations) included artists renowned for their fineness of feeling: John Ruskin (who extolled the virtues of chemistry), George Eliot (widely interested in chemistry, geology, and biology), Gerard Manley Hopkins (fascinated by the new physics, and a contributor to *Nature*). J.M.W. Turner positively reveled in the visual excitement and the sheer physical power of steam.

By the early twentieth century this smoldering enthusiasm consolidated into a widespread feeling that science could effectively be equated with rationalism, and that rationalism—and rationalism alone—could produce a better world: richer, healthier, and in the end more humane. World War I may have dampened enthusiasm for technology per se, but if anything, it reinforced the notion that human beings should, above all, *think*. They should not allow themselves to be driven simply by old-fashioned emotions, including patriotism. The new psychology (behaviorism on the one hand, Freud on the other), the new biology (genetics, biochemistry), the new philosophy (notably logical positivism, the belief that only the questions that could be answered with certainty were worth asking in the first place), the new art (outstandingly cubism) and architecture (soon to culminate in Mies van der

Rohe and Le Corbusier) combined to produce what Aldous Huxley, quoting Miranda from Shakespeare's *Tempest,* called the "brave new world." Eugenics might be harsh in some of its manifestations, but the human species needed to be purged with the same brisk, no-nonsense brush that was sweeping through the muddled academic corridors of science, philosophy, religion, and art. In short, eugenics belonged absolutely to the zeitgeist of the early twentieth century. Neither science nor politics was the leader of events. Eugenics did not belong exclusively to science, nor to the left or the right wing of politics.

I suggest, too, that in Britain there was a particular reason for believing that the human species was indeed producing an "underclass." In 1900, Britain found it shockingly hard to raise a convincing army to fight the Boers in South Africa. The men who came forward were tiny, sunken-chested, and often tubercular. This merely confirmed what was all too obvious on the streets: that the "lower classes" consisted largely of runts. These people were not necessarily "feebleminded," of course. Some of the greatest intellects and noblest spirits have lodged in small frames, from Alexander Pope to Mahatma Gandhi. But the education of late Victorian working people was perfunctory at best, and they were for many reasons dispirited. It is easy to mistake lack of confidence for stupidity. Truly, it seemed that these diminutive, downcast people were an underclass. Yet they seemed to have lots of children, which was not generally true, but their children were on the streets and so were conspicuous. The human race—or more specifically, the British people—seemed to be "degenerating," a much-favored term in those times. For economic reasons alone, humanity in general and Britain in particular needed to smarten up their act. Or so it was felt.

Wasn't it obvious, though, that the lack of stature and of spirit had little or nothing to do with inheritance—that they resulted simply from deprivation? Well, for curious reasons of biology, and without wishing to excuse "man's inhumanity to man," it could not have been as obvious as all that.

A person's height, at maturity, is of course influenced by his or her genes. The Masai people are genetically predisposed to be taller than the average Eskimo; Sudanese on balance are longer and skinnier than native Tyroleans. But several different environmental forces also influence final adult height. Nutrition during childhood has a huge influence, of course. Nutrition in utero also has an effect, although fetuses are very efficient parasites, even of malnourished mothers. Nonetheless, they cannot procure what the mother cannot provide, so maternal nutrition matters too. Much less obviously, adult height is related to height (length) at birth (assuming the birth is not premature): small babies *in general* make small adults.

Even less obviously, the length of a baby at birth is related to the mother's own length when she was born. It is as if mothers have a physiological memory of their own gestation and use this memory to judge what size of baby it is realistic for them to give birth to. The mechanism is subtle, but when you think about it, it is necessary, or at least very useful. In general, it pays a mother to produce a big baby, for big babies fare better at birth (leaving aside those that are large for reasons of pathology). She should, then, put as much nourishment into the fetus as possible. But if she puts in too much, she herself will die and so, then, will the baby. It is good, therefore, to be able to *adjust* the amount of nourishment provided, according to the prevailing circumstances, for in nature circumstances vary enormously. How should her body judge the appropriate input? What better than to remember, at the physiological level, her own prenatal experience?

Thus a person's height is influenced by his or her *mother's* experiences as well as his or her own, and the mother's experiences, in turn, reflect her mother's experiences, and so on. Undersize children who are well fed will of course grow bigger than they otherwise would. But they may not grow as big as their genes would theoretically allow them to, because their mothers were (probably) deprived as well. Thus it takes several generations of good feeding to restore an undersize population to its "proper" height, or at least to the height that its genes would allow.

This mechanism *looks* Lamarckian, since the experiences of the mother (and grandmother, and great-grandmother) are apparently inherited by the next generation. But it is not. The genes are not affected by the nutritional status of the lineage. It's just that genes are subtle, as this book has emphasized: they provide mechanisms that can respond to change. But the genes themselves have been selected by Darwinian means.

Be that as it may, the late Victorian "upper" classes perceived that the "lower" classes—the third- and fourth-generation offspring of the industrial revolution—were small and meek. Furthermore, their smallness and meekness were passed on from generation to generation. It *looked* as if the runtishness and lack of spirit were truly heritable, in the sense of being brought about by defective genes. So the *idea* that the British people (and the human race, by extrapolation) could "degenerate" was not quite as perversely stupid as it now seems. In truth, diminutiveness and dispiritedness merely revealed the flexibility of the genome. But it did not look that way at the time.

Even in World War I, Britain was obliged to put together "bantam regiments" of small men, who often acquitted themselves extremely well. Now such runtishness is rare. In the end, Britain cleaned itself up socially rather than genetically, and its people's physical improvement really can be seen as a triumph of the welfare state. But because it takes several generations to outgrow the nutritional shortcomings of earlier generations, Britain's people—indeed the people of much of the world—are still becoming taller, generation by generation. The poor prostitutes who fell victim to Jack the Ripper, the notorious serial killer of late Victorian London, were mostly around four feet tall. One of them was nicknamed Tall Lizzie. She was a towering four feet six. Modern girls are typically at least a foot taller than this, and girly sixfooters are already becoming commonplace. I am nearly three inches taller than my father was; my son is nearly five inches taller than I. There is no point dreaming of a career in basketball these days if you are much under six feet six.

How tall will human beings become, as successive generations continue on a high nutritional plane? For how many generations will the current trend continue? What would happen if people who are genetically predisposed to be very tall—like the Masai and the native Sudanese—were fed on Texan diets for four or five generations? Will tomorrow's basketball teams consist entirely of eight-footers?

We must wait and see. Meanwhile we may reflect that nature is usually more complicated than it looks, and that it is dangerous to be carried away by first impressions and novel theories. It would be naive (and just plain wrong) to suggest that the early eugenics enthusiasts were all bad people. But they show how easy it is to be misled.

Let's just ask a few biological questions, however. Let's assume for a minute that some regime did adopt a formal eugenics policy, setting out deliberately to "improve" people. Of course this hypothetical regime would first have to decide what it meant by "improvement." What would the human equivalent be of milk yield in cattle, or egg output in poultry? Well, perhaps the regime could produce different breeds for different purposes: athletes, musicians, and so on. But the quality that they would probably seize is intelligence. Specifically, then, would it be possible to breed for greater intelligence?

COULD WE BREED MORE INTELLIGENT PEOPLE IF WE WANTED TO?

The answer is surely yes, but (a) the problem is not quite as easy as it seems (or, apparently, as it seemed to Hitler's eugenicists), and (b) although we might readily raise the *mean* (roughly speaking, the average) IQ of the population, it is not obvious that we could improve the top end. That is, we might produce more people able to get A's at Princeton, but it might be very difficult indeed to produce anyone who was significantly more intelligent than, say, Niels Bohr. That is a guess, but it seems sensible. (Throughout this discussion I am assuming that the quality measured by IQ tests really does reflect what most people mean by "intelligence," but I am also defining "intelligence" broadly, to encompass all the many things people do that clearly reflect mental ability, from math to poetry.)

Little is known about the genes that influence intelligence most directly—as critics like to point out, there are no known "genes for intelligence"—but that need not stop us from making general observations, at least to get the ball rolling. Some scientists affect to dislike speculation, but if nobody speculated, there would be no science at all. Classical genetics had reached a high stage of sophistication and had become vital in agricultural breeding and in medicine before anyone even knew that genes are made of DNA, so it is not so damning as some commentators clearly imagine to point out that no *particular* "genes for intelligence" are known.

It is clear (on grounds both commonsensical and empirical) that many genes *affect* intelligence. In general, the presence and importance of particular genes is discovered only when they are polymorphic: that is, a range of alleles is possible at any one locus. Often, genes are discovered when mutant alleles cause obvious disease. Many of the 6,000 or so known alleles that can lead to single-gene disorders cause what textbooks describe as "mental retardation." We all know how sensitive our minds—including our ability to think—are to daily insults and stimulants such as fatigue, alcohol, nicotine, caffeine, fear, worry, and so on. Similarly, any gene that affects any aspect of body chemistry—which all of them do, to some extent—could in principle affect brain function. Biologists are wont to suggest, as a kind of informed guess, that around

10,000 of our 34,000 genes affect our mental ability. But we might almost suggest that virtually all of them probably have some kind of input, or would do so, if present in mutant form.

Clearly, it is hard to design breeding programs if there are 10,000 genes to play with all at once, but the picture is not totally hopeless. Beyond doubt, some of the genes are more directly involved than others. Those that produce particular neurotransmitters or help organize particular synapses are clearly among the most closely involved. It would not be at all surprising, then, if in time biologists discovered particular alleles, or combinations of alleles, that are closely associated with high IQ and that seem to have a particularly positive effect on IQ. Such alleles surely will be identified as geneticists map and sequence the genomes of more and more people.

Even before such mapping is done we can see in principle—just by applying the basic ideas of classical genetics—that human beings would be more likely to have a high IQ if they possessed a greater number of the alleles that most directly boost IQ, and fewer of those that depress it. That is obvious, and the principle applies whether we are talking about just one allele that enhances IQ, or half a dozen, or a thousand, or 10,000.

The broad problem, then, is to produce a population whose gene pool contains as many of the helpful alleles as possible, to ensure that each individual within the population has the greatest chance of inheriting a good proportion of those alleles. So how, in general, is this done? By the two basic methods discussed in chapter 8, mainly in the context of plants: crossing and selection, and mass selection.

The time-honored method of crossing and selecting is harder than it looks, especially when the required character is polygenic (or has various components, *each* of which is polygenic). An individual with good looks must obviously possess a fair proportion of the desirable alleles. But his or her fineness of feature also depends critically on the particular *combination* of those alleles, and the combinations are broken up and rearranged during sexual reproduction. For this reason alone (and there are many others), the offspring of handsome fathers and beautiful mothers can be very disappointing. Nineteenth-century breeders expressed this phenomenon as "reversion to the mean": the offspring of outstanding parents tend to revert to the average level of the population. Sometimes you may strike it lucky, as suggested by my anecdote of the charming and handsome German born in a Nazi breeding program. But in general you cannot just "cross the best with the best, and hope for the best." Furthermore, we saw that plant breeders typically cross tens of thousands of individuals from the parent generation to produce millions of offspring, from which they may select only one in a thousand. Such an exercise in human beings would require a continent, and require 20 years per generation. Crossing and selection, then, seems a bad general strategy for breeding more intelligent human beings by classical methods.

Mass selection is more appropriate for features that are highly polygenic (or for combinations of features) and in species that are naturally outbreeding. At present—by definition—the mean IQ of human beings is taken to be 100, with the vast majority being between 80 and 120. We could beyond doubt raise that mean IQ, probably significantly and within a few generations, simply by ensuring that people were not allowed to breed unless they had an IQ above, say, 95. The political control would be draconian and undoubtedly hideous, and many fine qualities might be lost by such a strategy. But it would surely work. Mass selection involves random breeding and is clearly very different in strategy from the Nazi attempt to produce better people by crossing particular individuals. The Nazis' mass elimination of large groups clearly had nothing to do with any concerted attempt to raise IQ. True, they attacked the people they diagnosed to be feebleminded (although we might question the diagnosis). But they attacked people on racial grounds with even greater vigor, with no regard at all for their measured intelligence.

But although mass selection would probably produce a greater proportion of clever people, it is hard to see how it could produce super-Einsteins or super-Shakespeares. The problem lies with the long-understood issue of heterosis. Some alleles do not produce their best results unless they are matched, on the equivalent locus on the homologous chromosome, by a different allele. As a top-of-the-head calculation we might reasonably speculate that people have the potential to be geniuses when they possess, say, alleles at 5,000 loci that enhance intelligence (and very few that detract from intelligence) and when, say, 100 of the 5,000 are heterozygous—with each heterozygous locus requiring *particular* pairs of alleles. When excellence requires heterosis, and when there has to be heterozygosity at several or many different loci, no breeder can guarantee to produce, or reproduce, the quality required. Individuals that possess such combinations of alleles are unique, never to be repeated—at least by the methods of classical breeding. This is why breeders of F1 hybrids—which are highly heterozygous—commonly begin with parent populations that actually look rather feeble. The parents are feeble because they are highly homozygous, but if they were not highly homozygous, they could not produce heterozygous offspring to a reliable specification.

It is fun (harmless fun, I suggest) to apply these simple plant-breeding principles to what we know of human beings. High ability, in many areas, undoubtedly runs in families. There was a dynasty of musical Bachs; there are many dynasties of painters (the Brueghels, the Teniers), and even more of architects and engineers (Barry, Pugin, Nash, Brunel, and so on). Bragg senior and junior both won Nobel Prizes, and the grandsons of Thomas Henry Huxley included Aldous, Sir Julian, and Andrew, who became president of England's Royal Society. The Medicis produced more than their share of outstanding individuals (with my vote going to Piero the Fatuous, who famously commissioned Michelangelo to build him a snowman). The Brontë sisters are the most famous siblings of literature. It is clear that high ability can run in families, although it seems obvious that environment and opportunity are just as significant as genetic input.

But although outstanding ability can run in families, it is rare for that extraordinary quality we call genius to do so. Only one of the Brueghels (Pieter the Elder) and one of the Bachs (J. S.) are generally taken to qualify. Jane Austen and George Eliot have commonly been called geniuses, but the term is rarely applied to the Brontës, talented though they were. If genius really does depend on a particular combination of particular alleles heterozygously represented, then this is entirely unsurprising. Neither should we be surprised if genius sometimes turns up in the midst of families that seem to have no outstanding talent at all. Beethoven is commonly presented as the prime example. Newton is another.

Some draconian regime could almost certainly raise general IQ—by mass selection, rather than by particular crosses—but it would probably have to wait for genius to turn up randomly, just as human societies have always done. Much more to the point, we should learn to treat genius much better than we do. The geniuses that have died before their time because of society's maltreatment make a truly horrendous catalog (which of course includes Mozart, the twentieth-century composer Samuel Coleridge Taylor, and Alan Turing, perhaps the most extraordinary scientific talent of the twentieth century, who committed suicide after harassment by the British government).

Overall, then, it would clearly be very difficult to make significant changes in human beings by traditional breeding methods even if we did introduce draconian policies. The logistics would still overwhelm us. Of late, however, many have claimed that genetic engineering really could produce dramatic changes, and indeed that it is bound to do so. The "designer baby"—structured genetically to specifications—is sometimes said to be just around the corner. Again, leaving aside matters of desirability, how realistic are such claims?

WILL WE EVER SEE "DESIGNER BABIES"?

Genetic engineering is qualitatively different from conventional breeding, although this has often been denied of late. Conventional breeders cross one individual with another to try to combine the desirable alleles of both in the offspring. Too bad if one of the individuals has only one desirable allele—thousands of other alleles are brought along as well and generally have to be got rid of by subsequent rounds of backcrossing. But genetic engineers are precise. They pick out the one gene (or the few genes) they want and add them ad hoc, or they take away or change individual genes. This is a conceptually distinct operation.

A "designer baby" is one that may have had many genes added by these ad hoc methods, to produce a human being to a particular specification. Technically, in theory, this must be possible. In principle, it is becoming possible to create genomes with any specified sequence of genes. It surely will be possible, then, to create a sequence of genes that will produce human beings who can do whatever their designers require of them: anything, that is, that does not defy the laws of physics. Neither should we second-guess what might be physically possible. We might guess that a basketball player ten feet tall would be mechanically unstable and suffer disastrously from back problems, or faint when he bent down to touch his toes. But don't hold your breath. Such disasters are plausible, but not predictable from first principles.

It would probably prove very difficult to produce supergeniuses merely by conventional breeding, even though we might readily raise the average IQ. But by genetic engineering it might indeed be possible to create supergeniuses (with a measured IQ of, say, 400). This presumably would require geneticists to identify the genes that most directly influence IQ, explore the way they operate and the combinations in which they work most effectively, and then add more that are similar or work even better, and so on.

My point is not that such outcomes are inevitable, and certainly not that we *should* go down such routes, but that it is dangerous to discount them a priori. Some scientists in recent years, alarmed by their own pending power, have sought to ward off public criticism by suggesting that the various "nightmare scenarios" are just not possible. Such a stance, to put the matter plainly, is dishonest. Outlandish scenarios *are* in principle possible. If we want to remain in control of our own lives and societies, it is sensible to spell out the possibilities as clearly as we can and then ask what we are going to do about them: whether those possibilities can be used for benign purposes (and what we mean by "benign") and how to avoid courses that are undesirable.

However, there can be a large gap between what is theoretically *possible,* and what is in practice feasible. I do think "designer babies" are possible, but I cannot see that they are feasible. That is, I cannot envisage a technically or politically plausible series of steps by which we might produce them. Over the next few decades we probably will see some forays into human genetic engineering, but these will always be of an ad hoc and highly circumscribed nature. We will not slide along the hypothetical "slippery slope" toward the full-blown redesigned human being, simply because there is no such slope. Instead, there is a severe obstacle course. From an ethical and a political point of view, the obstacles that can already be envisaged are a good thing. They will serve as milestones, and we can stop at any one of them.

TRANSITIONS, OBSTACLES, AND MILESTONES

We have already identified most of the principle milestones on the road to the designer baby. There is a conceptual difference between therapy—the alleviation of obvious suffering—and the attempt to improve on a system that is already functioning well enough. In some cases it is hard to identify the borderline, but we recognize the broad distinction well enough. In general it is the difference between medicine, intended to cure discrete pathologies, and tonics, designed to gild the lily. In Western medicine we tend to approve of the former and take a puritanical view of the latter. The same principle can be applied to genetic interventions. We have also distinguished between ad hoc interventions, intended to help particular individuals during their own lives, and germline therapy, which is intended to correct, or rather prevent, specific genetic disorders in future generations. Eugenics deliberations reveal the difference between attempts simply to ensure that future generations are like *us* and attempts to push the human species toward some preenvisioned goal. As long as we can identify such distinctions, we can frame laws, or at least define principles, that in theory should enable us to call a halt at any particular point.

There are some difficult cases, however. One is offered by all those alleles that do not lead to single-gene disorders like Tay-Sachs and cystic fibrosis but that *do* predispose to particular damaging conditions, such as diabetes and coronary heart disease. Modern convention has it that we are liable to suffer from coronary heart disease if we consume large amounts of saturated fat. True, but in fact only some people get heart disease on a high-fat diet. Others can eat lard all day and escape. The sufferers are the ones that possess particular genes that predispose them to arterial damage when the arteries are exposed to large amounts of fat. Similarly, we can find environmental causes of lung cancer, breast cancer, diabetes, emphysema, and any disease you may care to mention. But you will always find some genetic predisposition as well. Should we accept that in the modern world we *are* subjected to environmental pollutants and exposed to high-fat diets, and regard those predisposing genes as maladaptive? If so, should they become targets for gene therapy?

My own immediate answer is "No, they should not," for all kinds of reasons. Part of the reason is puritanical: it is incumbent upon us to live sensibly, if we have that option, and not simply to reach for the technological "fix." My greater reason, however, is common sense and some knowledge of biology. We have seen the principle in the case of the sickle-cell anemia allele. On the face of things, it looks all bad; but in reality, in context, the allele in a single dose saves a great many lives. The same is surely true of the many alleles that seem to predispose us to cancer, heart disease, or other maladies. True, in the context of modern, perhaps too-affluent societies, they seem to be bad for us. But in other contexts they might be very good. Besides, *most* genes are pleiotropic, and the gene that seems to predispose us to diabetes might also have a great many effects that are extremely beneficial, even in our present environment. The grand generalization applies, too. It is one thing to identify a single gene that is obviously life-threatening and try to do something about it. But we should not take liberties. We do not know enough—and never will—to be able to fiddle with genes without risk. If the risks are small but the dangers are obvious, even if they remain theoretical, then leave well enough alone.

Still, though, we can envisage test cases that must give pause for thought. In *Remaking Eden,* Lee Silver offers a case for germline therapy. Suppose, he says, it were possible to add a gene that would confer protection against AIDS. Would this be justified? Of course, you might argue that AIDS can generally be avoided by behavioral changes and so is not a good candidate. What then of malaria, which kills over one million people a year and has so far avoided all attempts at vaccination, and for which there are no fully satisfactory drugs? What of TB, which is currently attacking humanity afresh? Both diseases in principle *might* be prevented by adding or subtracting the appropriate alleles. It would be surprising if this were not the case. The malaria parasite, after all, must recognize the red blood cells in which it lodges by their surface properties, and subtle alteration could put the parasites completely off the scent. Germline gene therapy *might* prove the best and cheapest protection against malaria. Could anyone then suggest that it was too equivocal *in principle* not to be used?

Such cases are genuinely tricky. Perhaps they do serve to show that we should not dismiss germline therapy out of hand. But even so, such interventions would be ad hoc: specific defenses against specific diseases that we know can be particularly nasty and intractable in particular environments. The idea of the designer baby is generally broader than this: it implies greater height, higher intelligence, greater beauty. Again, leaving ethics aside for a moment, *could* we enhance such qualities by genetic engineering?

In principle again, the answer must be yes; but again, it is very hard to see how such attempts could ever be made. One obvious issue—only one, but it makes the point—is safety. Qualities such as intelligence and beauty—and even height—are polygenic. This does not mean that they could not be affected by the addition or subtraction of single genes, for they certainly could. So too, by crude analogy, the performance of an engine might be enhanced by introducing any one individual part of higher quality. However, the single genes that are introduced would have to work harmoniously with all the other genes that are already in place. There can be no guarantees that they would do so. Adding an extra gene that you think *might* fit the bill might upset the apple cart instead. Genes that promote the output of growth hormone produced monsters when introduced into farm livestock. Future manipulators of human beings would be more sophisticated, but the principle still applies.

Then again, *any* gene is liable to prove pleiotropic in its effects. When single-gene disorders are corrected, a defective allele is replaced by the normal version; we know what the normal one does by observation of healthy people. If we added novel genes to people to enhance height or intelligence, we could *not* predict precisely what other effects those genes might have. Nature cannot be second-guessed. We would just have to wait and see. This principle also applies, of course, to any attempts to try to provide protection against malaria or TB by means of germline therapy: but at least in such cases the aim would be to save life, and we can argue that the goal outweighs the risk. But when the goal is simply to add a few points of IQ to give some economic advantage, we might reasonably feel that all risk is unjustified.

In general, we might suggest that to replace deleterious alleles with normal ones is like proofreading: correcting the obvious spelling mistakes and removing the blots. But to enhance a genome that already operates reasonably well is an exercise in serious editing: like changing the end of *King Lear,* for example, as it was once fashionable to do because the death of Cordelia was considered too sad. The Human Genome Project will give us a printout of the human genome that will be analogous to a dictionary: at least we will have listed all the words in the language. Actually, though, it won't really be a dictionary, since it will take some time to work out what all the words actually *mean.* But even if it were a finished dictionary, even of the finest quality, would it really enable us to *edit* the genome? For if genes are analogous to words, then the genome as a whole is analogous to literature. There are cross-references, puns, allusions, colloquialisms, dialects, redundancies, and nuances of all kinds. Would you attempt to rewrite *King Lear*—or to edit a medieval poem in Chinese—if all you had was a dictionary? Would any sane person do this?

Then there are further, technical points. All technical difficulties can in principle be overcome, but again, there is a clear difference between what is technically possible and what is feasible. Germline genetic engineering in animals depends on the technology of cloning, by cell culture and nuclear transfer. This technology raises its own hazards. Nearly 300 embryos were required to produce Dolly. Of course, the techniques will become incomparably better as the next centuries unfold, but cloning will still be more risky than the standard method of reproduction. Furthermore, some of the failures on the route to Dolly and the other cloned sheep ended in late abortion, deformity, and neonatal death. In human medicine we tolerate early abortion—nature itself, after all, is incredibly prodigal in the early stages of pregnancy—but late abortion is still most disturbing, and deformity and neonatal death are simply horrible. If any attempt to produce a "designer baby" ended in just one such accident, the whole endeavor would surely be called off. Yet it is hard to see how accidents could be avoided. No procedure in medicine or any other field of human endeavor ever leaped straight to perfection. But anyone who offered technology to produce designer babies would effectively have to promise perfection from the outset.

No one sane would attempt such feats in humans until the techniques had been polished and honed in other species. That at least is the best scenario, but in practice, guns tend to be jumped. But even if the experimenters worked conscientiously through the phylogenetic tree—marmosets, macaques, chimpanzees—there would still be a leap of species at the end, and procedures that seem perfectly safe in other species *might* have ill effects in humans. No one can say a priori that that would not be the case. Besides, many people—including me—would question whether it is justified to carry out hazardous procedures in other primates merely to further the dubious ambitions of those who seek to produce extratall or extraintelligent humans. Here, then, is another obvious point before which we may draw our ethical lines.

Then of course there is the general point, made elsewhere in this book, that although we may develop procedures that achieve what we want them to achieve, and although those procedures might be underpinned by convincing and internally consistent theory, we can never *know* that we know everything that needs to be known. It is logically impossible to possess complete knowledge, or at least, to know that we possess it. So however many experiments we did en route to the designer baby, and however deep and intricate our theory became, our knowledge would always be imperfect. We would always be operating, to some extent, on a wing and a prayer. That is the general condition of humankind. All physicians and engineers must operate and build in the absence of perfect knowledge. If the prize is obvious—to save life—and the risk is calculable, then the adventure may well be justified. If this were not the case, none of us could ever take an aspirin, and no one would ever build a bridge. We know, however, that we sometimes come unstuck even when the science seems well understood and there are a thousand precedents. Every now and again, bridges built to impeccable standards and irrefutable theory fall down. When the rewards are highly dubious, and the risks are all too easy to envisage and are potentially horrendous, then surely we should back away. The point is obvious. There is no slippery slope to the designer baby, simply because in practice there is no easy, logical, step-by-step passage from the conceptually simple correction of single-gene disorders to the totally prescribed genome.

A final technical point: in 100 years' time the very notion of the designer baby might simply seem rather quaint, like most science of the early twenty-first century. The alternative route is through pharmacology. More and more, modern drugs operate at the level of the DNA. In a few generations' time we will surely have drugs that can control the expression of genes to the nth degree. Why change a gene permanently and irrevocably if we can simply add some agent that will invoke what is required from existing genes, or mimic the function of genes that are not there?

But of course, prediction in science is hazardous. I cannot see how designer babies will become reality, but neither do I claim to be omniscient. The *idea* of the designer baby is now on the agenda of humankind, and until science itself comes to an end or human beings re-evolve along nonintelligent lines, it will remain there forever. "Genetic determinism" is just a slogan for the lazy-minded, but it remains the fact that if we redesign the human genome, then ipso facto we will be redefining what we mean by *humanity*. That would be the most extraordinary, the boldest, and the most threatening endeavor human beings have ever undertaken: to redesign ourselves, declaring our present selves obsolete.

I do hope we do not do this. We might reasonably leave the last word to Marvin Minsky, of the Massachusetts Institute of Technology. He has pointed out that we do not give enough credit to our present selves. He suggests that if you really want to appreciate something, you should try to make a machine that will do the same things. He and his colleagues at MIT have contrived these past few decades to emulate the human brain, and have been chastened, staggered by the subtlety and intricacy of the real thing. We extol what we call genius to the heavens, says Minsky— Mozart, Einstein, Shakespeare. Yet the difference between any of these paragons and, say, a coat-check attendant is too small to measure compared with the difference between the attendant and, say, an earthworm. Yet an earthworm is not an inconsiderable creature; it is many, many times more complicated than, say, a protozoan. It is actually easy to program computers to make music of commendable intricacy, and they do blinding math. Poetry and general creativeness come harder, but the day will surely arrive when we will not be able to tell if some new sonnet is the work of a person or a machine. By contrast, it would take more computer power than that of the Starship *Enterprise* to program a robot that could hang up coats, hand out tickets, and exchange pleasantries while plotting a career in the movies. Such extraordinariness we take for granted: just sitting on a chair, or holding simple conversations, and adjusting style and content according to who is being spoken to (boss, friend, client, grandmother, little daughter). Yet these abilities beggar belief. "What a piece of work is a man!" exclaimed Hamlet.

And he was absolutely right. For reasons of biology alone— leaving aside all issues of ethics and theology for a moment—we should surely not tinker lightly with such a creature. The idea that we can improve on our own good selves by fiddling with our genes like washing-machine repairmen seems to me not simply hubristic, as I will suggest in the epilogue, but ludicrous. It has taken us 5 million years to evolve to our present state. Natural selection really does work and has been a hard taskmaster. It is obviously dangerous to override it. We should remember that our apparent successes as breeders of livestock are extremely limited: domestic animals are bred as one-dimensional creatures that give ridiculously large amounts of milk or lay absurd quantities of eggs. The qualities their ancestors enjoyed as wild animals may simply be squandered. But we cannot afford to squander the thousands of subtleties that our evolution has built into ourselves, and in particular into our psyches. We should accept ourselves, as Oliver Cromwell was content to accept himself, "warts and all." Or at least, we might contrive to preempt the suffering that is caused by the most objectionable warts, the obviously deleterious alleles. Apart from that, we should reflect how nobly the genetically pristine human being can perform when given the chance—three square meals a day, freedom from disease, social harmony, education—and focus, as all sensible social reformers have always done, on improving our environment. The rest is rather distasteful and, in the end, silly.

But whether we like it or not, the human clone and the designer baby, the reinvented human being, will stay on humanity's agenda for as long as science itself is practiced. With such power before us, we have to ask as a matter of urgency, what is *right* for us to do. Some have suggested that these new technologies raise no "new" ethical issues, a point that largely depends on what is meant by *new*. They certainly raise the ethical ante. After all, we cannot be held morally responsible for events that we cannot control, but we are answerable for those that we do control. In the normal course of events, we cannot control the genetic makeup of our offspring. We do have *some* influence, because we choose our mates carefully, but the process of genetic recombination during the formation of eggs and sperm ensures that the genetic details of our offspring are not ours to specify. But if we clone children, or engineer their genes, then we are *prescribing* their genome. Our responsibility then, for all that befalls them, far outstrips that of any parent. Noblesse oblige. It is too casual by far to say there are no new issues. We must look deeper.

Epilogue:

What Should We Do

With All This Power?

This book is concerned with fundamentals. I have sought to show how all the complexities of modern genetics, and all its ramifications, flow naturally from Mendel's initial notions and his rows of peas: those notions, plus the chemistry of DNA and the evolutionary insights of Charles Darwin. Most of this book has been concerned with the facts of the case: with what *is*. In this epilogue I want to ask what *ought* to be. But I want to apply the same general approach: using reverse engineering, working back from the complex surface of things to track down the fundamental principles that underpin—or ought to underpin—ethical and economic discussions. As with genetics itself, I feel that the fundamental moral principles are in the end simple. In ethics, great truths may indeed spring from the mouths of babes and sucklings. Beware the rhetoricians and arm wavers. As Mendel himself so abundantly showed, simplicity should not be confused with crudity, or complexity with profundity.

WHY SHOULD WE BE CONCERNED?

If creatures have no consciousness and no power to change the world around them, then issues of ethics do not arise. Cows or lions are not morally accountable. *Forgive them, for they know not what they do,* and even if they did, they could not, in practice, make much of an impact on the world. Human beings, however, *are* conscious. We can decide what we want to do, or at least, if choice is an illusion, it is one that we do well to cultivate. From the time we first became human, we surely have felt ourselves to be morally accountable, at least to each other and perhaps (who knows?) to some deity. Furthermore, we have huge technical power. We really can change the world around us, do so restlessly, and have already done so from one pole to the other. Our power, like all power, is both destructive and creative. Destruction is much easier, and of course destruction and creation overlap. The things we build—farms, cities, domestic crops, and livestock—largely destroy what was there before.

Our control of the physical world has long been obvious: rapid transport, dams and highways, skyscrapers, and now, spectacularly, electronics, which has already brought us virtual reality and the Internet. So far, our control of living systems has been more tenuous. Modern medicine and agriculture are fabulous by the standards of earlier centuries, but in truth their successes are ad hoc. We do what we do, and we think we understand what we do. Yet in truth we have been as Newton said of himself: children playing with pebbles on the beach.

But this state of affairs is clearly changing. Biotechnologists are already at work on "designer crops." The present wave of GMOs—genetically modified organisms—that have caused such a fuss is only the beginning. It is not too soon to contemplate the "designer baby," even though the reality (if it is ever realized at all) must lie many decades in the future. Indeed, we might now sensibly consider the creation of life itself—not simply by modifying present life-forms, or rearranging them in the way that enthusiasts build custom cars, but from scratch, with the ingredients you might once have found (in less safety-conscious days) in a child's chemistry set. As late as the 1980s, the outstanding German biologist Davor Solter wrote in *Nature* that cloning of mammals by simple nuclear transfer was "biologically impossible." Never mind the particular issue that he was addressing. The point is that after Megan and Morag, Dolly, and Polly, the expression "biologically impossible" seems to have lost all meaning. We might reasonably contemplate *any* feat of biotechnology that does not break what Sir Peter Medawar called "the bedrock laws of physics," or defy the rules of logic.

We are beginning to acquire a power over living systems that will seem, as the next centuries unfold, to be absolute. It is an ancient principle, made explicit in feudal societies as noblesse oblige, that those who have power are obliged to act responsibly, and moral responsibility grows in direct proportion to their power. Absolute power implies absolute responsibility. We really do have the fate of our own and all other species on Earth in our hands, or soon will have. Morally, there is nowhere to run.

We should, however, add two conditional clauses, though neither reduces the burden of our responsibility, and one of them, if anything, adds to it. First, when commentators (scientists and otherwise) consider the future of biotechnologies, they typically underestimate the time scale. Thus experts have been predicting of late (not least on peak-hour British television) that in 20 years' time human reproduction will be mediated in vitro and sex will be a purely social pursuit. What nonsense. As I will argue later, there is good reason to suppose that sex will *always* be the preferred method of reproduction and that in vitro methods will always be reserved for special cases. Even if this were not so, 20 years is an absurdly short period for such a prediction. Science and technology seem to progress at a bewildering pace. But the universe is big and complicated and there are a lot of problems to get through, so complex technologies in reality take decades or centuries to unfold, especially when, as in this case, they are concerned with the biology of human beings.

Consider the history of genetic engineering. DNA was first discovered in the late 1860s, yet it was another 80 years before its function was properly appreciated, and another decade after that before its three-dimensional structure was worked out. Gene transfer was first broached in the early 1970s but did not become a truly practical proposition in animals until the late 1990s, when Roslin developed cloning technology. Thus, 130 years elapsed from the discovery of DNA to the manipulation of DNA in animals. Similarly, Edward Jenner initiated vaccination (using cowpox to immunize against smallpox) in the 1790s, and now, 200 years later—despite many spectacular triumphs, including the total elimination of smallpox—vaccines are still as hot a topic for research as ever, and still posing problems. As knowledge of the Universe expands, so do the mysteries and perceived opportunities. The same principles will apply to in vitro reproduction and designer baby technologies. It will be many decades before they come on line, and they will still be posing problems in two centuries' time. That, at least, is a more realistic time scale than two decades. Scientists exaggerate the speed of progress because they need to attract government grants and venture capital, for few would invest in technologies that might easily take a century to yield dividends.

On the other hand, the human species is not—as people seem so often to think—facing Armageddon. The world is not about to end, or at least, we have no reason to suppose that it is. Our descendants will still be here in 200 years, and indeed, given that human beings are living longer, those descendants could include the great-grandchildren of at least some of this book's readers. Our descendants will still be here in 500 years, or 1,000, or 10,000, or 100,000. I suggested in *The Day Before Yesterday* that in thinking about wildlife conservation, a *million* years is a sensible unit of political time. So yes, technologies take a long time to unfold. But there is an awful lot of time for them to do so. Present-day technologies may be promising more than they can yet deliver, but they will remain on humanity's agenda *forever,* and as the decades and centuries go by, the hype will be superseded, more and more, by reality. So the time scale for the future is far longer than is commonly presented, but the future will arrive nonetheless.

The second conditional clause is truly a caveat. As time passes, our control of living processes will *seem* absolute; but *seem* is the operative word. Our descendants will be able to do anything they may care to conceive, including the creation of novel life and even of intelligent novel life. But their *understanding* can never be absolute; that is a logical impossibility. Nature does what nature does. We are always condemned to trail along behind, trying to work out what is happening. We can never, as a matter of logic, be sure that we have not missed some vital component of the natural mechanisms. Because our understanding will never be absolute, we will, as the future unfolds, make mistakes, and every now and again some of those mistakes could be serious.

We can look at this issue in various ways, but let us take just one. A scientist may have a theory on how some aspect of the universe works. On the basis of that theory, a technologist may build some machine, or devise some procedure. If the machine works or the procedure succeeds, does this prove that the theory was correct?

In practice, we tend to take such success as proof, for what else have we got to go on? A theory that gives rise to a technology that works must surely be correct at least *in some respects,* and *as far as it goes.* Yet the theory may still be deeply flawed: wrong in highly significant respects, or simply inadequate. To be sure, the theory may be perfectly coherent and intellectually satisfying. It may be perfectly consistent internally, with all components following logically, one from another. But still, the theory may not match reality as closely as the observer may suppose.

We can demonstrate this by analogy. Ancient farmers typically surrounded their cultivations with rituals, sometimes vastly elaborate: sow your crops only in the light of the full moon; "ne'er cast a clout till May be out"; dance to make the rain come; and so on. *Some* of those ancient rituals make perfect sense when analyzed in the light of modern knowledge. Modern agriculturalists have increasing respect for traditional farming systems that once seemed simply to be steeped in superstition. British farmers, for example, are well advised to take account of late frosts, and not to "cast clouts" till May be out. Other rituals, however—like rain dances, perhaps—defy all analysis, and indeed seem to make no sense at all. Yet *all* the farmers' procedures derive from the same, coherent theory of how the Universe works: typically from some basic idea that the Universe is controlled by a deity who must be respected. But the fact that the theory is coherent, and *seems* to work when put to the test, does not mean that it is, in fact, correct. Some of the farming procedures that the underlying cosmology demands *are* helpful, but some are irrelevant, and some that could be helpful (like fertilizing the fields) are in fact being missed. The underlying theory may be coherent and satisfying, and the procedures that are based on it may work (up to a point), but the fact that they may work simply does not prove the overall validity or adequacy of the theory.

We can bring this discussion closer to home. A decade before Ian Wilmut and Keith Campbell produced Dolly by nuclear transfer and genomic reprogramming, several groups of scientists managed to clone various kinds of animals by various other means. All of them had a good idea of what they were doing, and why. All, in short, had coherent theories that seemed effective. In retrospect, however, it is clear that the pioneers of cloning often missed vital elements: that some of the procedures they carried out conscientiously were not, in fact, necessary at all, while other things they did inadvertently turned out to be crucial. It seems possible, for example, that in some of the early experiments in nuclear transfer, the receiving eggs were activated inadvertently before the nuclei were introduced just because they had been allowed to revert to room temperature. In such cases, the concentrations in the eggs of a substance known as maturation promoting factor, or MPF, were low at the time of nuclear transfer. If MPF had remained high, it would have been destructive. The scientists involved in these early experiments had no idea that MPF was significant in this context, or that it needed to be reduced. They just happened, in passing, to do things that proved helpful.

In short, in modern science as in ancient agriculture, the underlying theory, however coherent and satisfying, may not underpin the resulting technology as accurately as seems to be the case. So the success of the resulting technology does not *and cannot* demonstrate that the underlying theory is correct and adequate in every respect. The theory must have been good enough to succeed in the particular case that was observed, but if the conditions were changed somewhat, inadequacies could well be revealed.

The above is not a criticism of science or scientists. It is simply a point of logic, and as such, it is inescapable. As it has been in the past, so it will be in the future. Future technologists may "design" babies, and may even make new forms of life, and their successes will seem to justify the belief that their *understanding* is absolute, just as their control seems to be. But this will always be an illusion. Even if their understanding *were* complete, they could never *know* that it was complete. Always, there is likely to be more going on than the scientist is aware of. Always, life will pull surprises. As a final illustration, we may note that modern architects and engineers really do have deep and extensive knowledge of Newtonian mechanics and the behavior of every kind of material you might conceive. Yet when they build with novel techniques, their bridges and stadiums sometimes fall down (actually, they do so with distressing frequency). The failures demonstrate the principle that our understanding can never be absolute, and it is dangerous to assume that it can be. But bridges and stadiums are easy compared with the processes of life. We might look upon the failures, the twisted strands of metal, as Shelley's traveler looked upon the ruins of Ozymandias, and despair. At the very least, we should take serious heed.

Nevertheless, we can expect our more bullish descendants to behave *as if* their understanding were absolute, and the biotechnical feats they will pull will truly be wondrous. It will *seem* as if they can do anything. And if human beings can do anything they choose to do, then what *should* they do?

HOW DO WE JUDGE WHAT IS GOOD?

We may reasonably assume that the very first human beings were moral creatures, so we may likewise assume that the questions—what is good, and how do we know goodness when we see it?—are as old as humanity. I do not presume, in the next few thousand words, to provide the eternal verities that have eluded humankind for so long. I would like to offer a few observations, however.

First, speaking in a broad-brush way, the moral principles that philosophers have defined tend to fall into one of two main categories: either absolutist or consequentialist.

Absolutist arguments say, in effect, that morality is structured into the Universe. Such arguments are easiest to frame, and to understand, if we assume that the Universe is run by an omnipotent and omniscient God. Then we can simply say that what is "good" is simply what God says is good. What God says is *absolutely* right. Our only task as moral beings is to find out what it is that God requires of us. Obey God and you can't go wrong, even if God is asking you to sacrifice your son as a burnt offering, as Jehovah once demanded of Abraham—although, as it transpired, he was only testing Abraham's faith (Genesis 22:2).

Many philosophers, however, have tried to define absolute moral principles that should apply even if there is no God. These, again, seem to be of two main kinds. The first is exemplified—indeed, reached its zenith—in Immanuel Kant, in eighteenth-century

Germany. He strove to define absolute ethical principles a priori: principles that in effect have the same weight as the laws of science. Such principles should outweigh even the will of God; if God does not subscribe to those principles, then *God* is wrong. Kant called his principles categorical imperatives. This is not the place to argue whether Kant's imperatives achieve what is required of them (modern moral philosophers seem to conclude that his scheme contains flaws, perhaps inevitably) or whether his ambition was realistic. Let us just take it for the moment as a fact: that much moral philosophy has been concerned to define *absolute* principles of goodness and badness, even in the possible absence of a God who gives the orders.

The second main strand of nontheistic absolutist arguments takes as its premise that what is *natural* is "right," and what is *unnatural* is "wrong." Thus, to take a very simple example, it is natural for mothers to look after their babies, so this is self-evidently "right"; it is unnatural for them to shun their babies, and this is obviously "wrong." But many have pointed out, with various arguments, that we cannot derive "good" from "natural" so simply. The teachings of St. Paul illustrate the dilemma. He condemned sex between men and women out of wedlock *even though* it is "natural." We should, he said, exercise restraint. On the other hand, he condemned homosexual relations between men precisely because they were *un*natural. In general, he seemed to feel that unnatural was wrong, but that natural was not necessarily right. In the eighteenth-century the Scottish philosopher David Hume pithily pointed out, "'Is' is not 'ought'" (he never actually used this particular formula, but this is the usual modern précis), and in the early twentieth century the Cambridge philosopher G. E. Moore coined the expression "naturalistic fallacy." This expression summarizes and effectively dismisses *all* attempts to equate morality with naturalness. Biology, said Moore, is just as arbitrary a source of moral guidance as any other body of ideas we may care to invoke. Yet perhaps we should not dismiss the idea that good *equals* natural quite as peremptorily as Moore would have us do.

I suggest, as a psychological point, that whether or not we are serious students of moral philosophy—whether or not we are committed Catholics or Muslims or followers of Kant—most of us *feel*, in our bones, that there *are* absolute moral principles. There may be such principles out there in the Universe, or there may not be; but I suggest that most of us *feel* that there ought to be. It is hard to find anyone who has no sense at all of absolute good and bad. Even the most hardened members of the Mafia love their mothers, and feel such love to be sacred. The feeling that there is, somewhere, an absolute good is presumably what drove Kant and many a theologian to pursue their intellectual quest. Perhaps King Arthur's search for the Holy Grail is a metaphor for this. Whether or not we ever find this grail—that is, define the absolute principles of goodness and badness—the search for it is worthwhile. We should never stop seeking to define what we mean by *good* and *bad;* and the notion that there are *absolute* standards of goodness and badness is surely a useful heuristic, a guide to thought. If we lose sight of the notion that there might be moral absolutes, then, it seems to me, we could be in deep moral and biological trouble.

Consequentialist arguments, by contrast, judge the goodness or badness of an action by its outcome. Perhaps the most famous and straightforward form of consequentialism was "utilitarianism," first outlined in the late eighteenth century by the English philosopher Jeremy Bentham. Utilitarianism is often summarized, cavalierly but accurately enough, as "the greatest happiness of the greatest number." The implications are obvious: if an action makes somebody happy (albeit only by reducing his or her pain), then it is good, unless it makes more people unhappy in the process.

In practice, it can be very difficult to disentangle absolutist arguments from consequentialist arguments. For example, we might argue that an action is bad even though it makes somebody happy, if the action itself is "bad." Sadists are made happy by other people's pain. Is it right to torture people to keep sadists amused? In such a case, of course, the delight of the sadists is matched by the distress of the victim, but if there is only one victim and a hundred whooping lookers-on, isn't the criterion satisfied, of the greatest happiness of the greatest number? Bentham was of course aware of such criticisms—and there are many others, much more subtle—but still the point remains. Human happiness may be a guideline in our search for goodness and badness, but it does not *by itself* tell us all we need to know. Still, we feel there are good reasons and bad reasons to be happy, and that the goodness and badness of the reasons must be judged independently of their effect on particular people.

The prickly issues of human cloning contain the two threads of thought, the absolutist and the consequentialist, and show how they intertwine. Many have argued that cloning is justified if it helps infertile couples to have a baby of their own that is genetically related to at least one of them. There is no more powerful human instinct (the argument has it, accurately) than the desire for a child. If cloning is the only route to reproduction, as in some circumstances it might be, then who should deny such a demand?

The issue can be argued purely along consequentialist lines, and many have attempted this. Just to make sure we have covered all possibilities, we could lay out the potential gains and drawbacks in a Latin square, like the ones used to describe the matings of Mendel's peas. Along one axis we could list all the people involved in the cloning: the donor of the nucleus, the donor of the egg, the obstetricians and technicians, society at large, and so on. Along the other axis we could list the possible gains and losses. Some of the people involved would grow rich by the procedure, which might be seen as a plus, but some might suffer psychologically, like the surrogate mother who gives birth to the cloned baby, and this would obviously be a minus. Then again, as Ian Wilmut has often asked, what psychological traumas might the family suffer as they bring up a baby who is a genetic replica of one parent but not the other? We can certainly envisage some special difficulties, and these too must be listed among the minuses.

But the chief player in all this is the baby itself. If the baby were to be deformed or otherwise incapacitated, then obviously the procedure is unjustified. We must assume, though, that cloning would not be attempted at all—in animals or in humans—unless there was a reasonable chance of success. So what is a "reasonable" chance in the context of a human baby? Late abortions, deformities, and neonatal deaths of the kind that occurred in the attempts to clone sheep at Roslin would clearly be unacceptable. So would we suggest that a one in a hundred chance of disaster was acceptable? Or one in a thousand? In practice, it is foolish simply to pluck such figures out of the air. We might rather observe that *natural* births, generated by the time-honored sexual means, *sometimes* end in disaster. So perhaps we might suggest that cloning would be acceptable provided the risk (of late abortion, deformity, neonatal death, or some later disaster) was no greater than in natural births. That would be a harsh criterion indeed, however, and impossible to judge until a great many babies had been cloned and statistics were available. It is probably more sensible to take some less demanding yardstick. Should we perhaps suggest that the risk of cloning is acceptable if it is no greater than that of some roughly comparable procedure, such as IVF? Since IVF is already widely accepted, many would feel that this would at least be a sensible compromise.

All this—including the vital assessment of risk—is consequentialism. We are simply asking who might be made happy by the birth of a cloned baby, and who might be unhappy, and what is the risk of either outcome. The absolutist would go further. Absolutists would suggest that *even if* everyone concerned is made happier by the birth of a cloned baby, it is *still* wrong. The principle has been well expressed by Prince Charles, heir to the British throne, not in this context but in that of GMOs. Such intensive biological interventions, he said, trespass "into God's territory."

Many people in this secular age are appalled by such language. Nevertheless, I contend, many people who are not overtly religious and would not themselves speak of "God's territory" empathize with the prince's misgivings, and would, if pressed, acknowledge "God's territory" as an apt metaphor for their feeling. Cloning might make people happy, but it is *still* wrong.

We might draw a parallel with a bank robbery. The robbers might be very polite and benign. They might make tea for the staff, make them laugh, and help them complete their crossword puzzles. They might donate the stolen cash to the local hospital. Even if they spent it on Rolls Royces, which is the more usual course, they would still be doing good in this capitalist age: helping keep a worthy industry afloat. Nobody loses much: the bank is rich and insured, and its customers lose too little per head to bother about. Besides, crime itself provides endless opportunities for respectable employment: insurance companies, builders of safes and burglar alarms, security guards, police, lawyers, prison guards, and all who service all of the above—van drivers, secretaries, and so on. The elimination of crime would create a huge hole in the economies of all organized countries and destroy some of our most respected professions. Lest you feel that such an example is too fanciful, recall that President Nixon once justified industrial pollution on the grounds that it fostered the cleanup industry. Yet we feel that bank robbery is wrong. The ends simply do not justify the means.

Cloning is not, of course, directly comparable to bank robbery. It is not conceived, as bank robbery is, as an offense against society. It is, we may concede, intended to do good. Yet the point remains. The resulting happiness or otherwise of the participants is not the only issue.

In short, whether or not we feel ourselves to be "religious," we all have some respect for absolutist arguments. Some people might not acknowledge this, but when appropriate cases are suggested (and I think the cheery bank robbery fits the bill), then even the most sceptical of consequentialists would concede the point, that *mere* human happiness is not the only criterion to be taken into account. There remain misgivings that may not literally emanate from some all-powerful deity, and may not be built into the fabric of the Universe, but must still be taken seriously. At least for the shorthand purposes of this argument, those misgivings can be taken to reflect a deep feeling, which all of us share, that in the end there *are* absolute criteria of good and evil. Even if that is not literally the case, we feel it to be the case, and what we feel to be the case cannot simply be ignored.

Finally, although in practice absolutist and consequentialist arguments are inescapably intertwined, we should at least try to keep them distinct *in our heads.* I have heard many an ethical committee at loggerheads, with some talking about human fertility and the pain of childlessness, and others talking about the will of God, and neither side connecting with the other; while the chairperson simply looks confused. In ethical arguments it is necessary both to propose moral principles and to ensure that all the arguments are neat and tidy, at least in structure, so that we know where we are. In actual ethical arguments, even in high places, one or both of these essentials are often lacking.

But let us speak more of absolutist arguments. We should ask, "What *are* the absolutist criteria of good and bad?" Then we should ask, "Where do those criteria come from?"

CRITERIA OF GOOD AND BAD: WHAT THEY ARE AND WHERE THEY COME FROM

Most of all, I like what David Hume said about ethics: that in the end, all our ethical sentiments are rooted in *feelings.* Our personal morality is driven by emotion. This is ineluctably the case. Moral philosophers write huge tracts, arguing this way and that, but different moral philosophers present different arguments, and of course make vastly different recommendations. In the end, inevitably and invariably, we find that among professional moral philosophers, just as among any group of idlers propping up the bar, some are right-wing and some are left-, some are authoritarian and some are liberal, some are softies and some are hard as nails. The pages and pages of argument merely serve to justify, albeit with somewhat more rigor than the average barroom lizard, a few basic predispositions and predilections. In fact, of course, the principal role of moral philosophers is *not* to tell us what is right or wrong. Their principal role is to take moral propositions—any propositions—and lay them out clearly for inspection. Their role is very like that of the lawyer, except that the lawyer is concerned specifically with the interpretation of law, whereas the moral philosopher seeks to make explicit the principles upon which laws can be based.

In a democracy *all* of the people ought to frame the initial propositions. Then we would hand them over to moral philosophers, who would argue them this way and that, just to see how the arguments pan out. If appropriate, they, in turn, might hand over the tidied-up arguments to government to become the basis of law; the laws would be worded, and later interpreted, by lawyers. That is how democracy *ought* to work, at least according to the interpretation of "democracy" I feel is common in Britain.

In reality, however, most of us don't have time to frame basic moral propositions. We have our own lives to get on with. Most societies have been content to leave the business of framing morality to experts, just as they delegate the teaching of children and the laying of drains. Who, though, are the appropriate experts?

Most people would say, if asked, moral philosophers, but this is *not* really their job. Their job is to take premises and argue them through, not to frame the premises in the first place. Some people (those who haven't had time to think it through) would perhaps say politicians. Politicians, after all, are often quite well educated, and they do in practice contrive to tell us all how to behave, so why shouldn't they put forward the basic propositions? Well, the main reason is not that politicians are outstandingly evil or amoral but simply that all politicians belong to parties and defend party positions: Social Democratic, Republican, Communist, and so on. As seekers after moral truth, we need to operate at a deeper level, to ask whether the bases of those parties are right or wrong. Some would put lawyers in charge of moral policy. This is not foolish: many lawyers clearly have a highly developed moral conscience, and most are clear-thinking, which is a sine qua non. But the criticism that applies to the moral philosophers applies, with interest, to the lawyers. It really is not their job to say what is right or wrong. It is their job to frame and interpret laws based on their society's notion of what is right or wrong.

In practice, however, ethical committees tend to be compounded primarily of moral philosophers, politicians, and lawyers, with scientists as expert witnesses and a token "layperson," commonly a housewife or a member of some minority. There is also, typically, some manner of cleric: a rabbi, a Catholic priest, a Methodist minister. Such a grouping is supposed to give us the rounded view. It is supposed to be the best we can do.

But the people I think we really need are prophets. Prophets are not the same as moral philosophers, although they may seem to drink from the same trough, and they are not the same as priests. It is the task of moral philosophers to discuss moral premises, but it is the task of prophets to put forward those premises in the first place. It is the task of priests to carry out the rituals and apply the teachings of particular religions, but it falls to prophets to frame the morality that underpins those religions. Of course a prophet may also be a moral philosopher (as was surely true of Christ) or a priest (like St. Francis) or even a politician (I would put Gandhi and Nelson Mandela in the category of prophets), but it is not necessarily so. Prophets are prophets. They define the deep premises.

Prophets are children of religion and manifestations of religion: Buddha, Moses, Isaiah, John the Baptist, Christ, St. Paul, Muhammad, and so on. You may feel, therefore, that in appealing to them I have sacrificed the attempt to discover deep principles that can serve all humanity. After all, isn't the religious approach to ethics to seek "the word of God"? What validity does this have if God does not exist? And why should an atheist, who does not believe in God, take note of God's alleged prophets? Besides, it is clear that different religions advocate different moral strategies. Morality based on religion thus becomes entirely arbitrary—a matter of whether we happen to be born into a Christian society, or a Muslim, or a Hindu. How can such particular points of view serve the needs of all humanity, or claim any fundamental, universal status?

Well, many people who consider themselves "religious" nonetheless reject much or virtually all of the theology that attaches to any one religion. Many modern Christians question whether the virgin birth of Christ, and his resurrection, are *literally* true. Though these two propositions are traditionally taken as the cornerstones of Christianity, some of the doubters even practice as priests. Many people believe, in short, that if you remove the theology from religions—including the central theological notion that God exists, and is the Creator of Heaven and Earth—then what is left is still worthwhile and is, in fact, indispensable, part of the fabric of being human.

So what do religions do *apart* from promulgate the notion that God exists (*pace* Buddhism, which is not theistic) and promote the idea that supernatural forces lie behind the superficialities that we perceive as "reality"? They do three things. First, they invariably seek to provide a complete narrative, a complete explanation of how the Universe works, how it came into being, and why it is as it is and contains the creatures it does, including us. The great religions seek to embrace *all* possible knowledge and experience. The attempt to provide a complete account seems to me to have some nobility, even if some moderns may doubt its possibility.

Second, all the great religions seek to frame systems of ethics. They all ask (or state) how human beings *ought* to behave, and they seek, furthermore, to weave those ethics into the grand narrative.

Third—and this is the part I feel is truly important in this context—all the great religions adopt the same kind of methods in reaching their ethical principles. None of them, as far as I can see, present moral *arguments* of the kind developed by moral philosophers and lawyers. Instead, the religions—through their prophets—simply *state* how we ought to behave. These statements are arrived at not by rational argument but by appeal to emotional response. The great religions, in short, *seek to cultivate and to define emotional response.* Religions have often been criticized for this approach, yet this is precisely what is needed. Emotion, as Hume said, lies at the root of all ethics. The *arguments* of moral philosophers that seem so supremely important are secondary. The underlying emotional response is what counts. As far as I can see, however, only religion seeks to refine emotional response in a formal way. Such refinement, I suggest, is the hallmark of religion. Those who seek to modify moral behavior by appealing directly to the emotions are employing the devices of religion, although they often do so in a disorderly fashion. (I feel, in short, that Hume's notion that ethics are rooted in emotion provides one of the principal justifications for religion; the cultivation of the emotional roots of ethical behavior is what makes it valuable even when it is stripped of specific theology. It is a pity Hume seemed to present himself as an atheist, and is generally perceived to be so.)

The ethical positions of the great prophets are statements of *attitude,* of emotional response. Of the great religions, this seems to be least true of Judaism, which is concerned primarily with law—with carrying out what is prescribed—and seems to make the least direct appeal to emotional response. Yet the commandments of Moses and the laws laid out in Leviticus do not simply advocate or proscribe particular actions. They demand particular attitudes: *honor* thy father and mother; *worship* the Lord thy God; *love* thy neighbor. Jesus, good Jew though he was, tipped the scales a little further, increasing the emotional demands—"Love thine *enemy!*"—and sometimes seeming to emphasize emotional response, namely compassion, at the expense of law, as when he effected cures on the Sabbath. Yet this was, in the end, just a shift of emphasis. The underlying emotional requirements are already implicit in Judaism.

All the great prophets, in all the great religions, arrived at their perceptions of attitude in the same kinds of way. Notably, they all spent time in contemplation, living ascetically but cultivating mental tranquillity. Long periods in the desert or in the mountains, far from the madding crowd, have been typical, if not de rigueur. Many modern leaders who might be classed as prophets, like Gandhi and Mandela, spent time in jail, in solitary confinement—that is, in metaphorical deserts. Solitude per se surely does not lead to moral enlightenment, but it is clearly part of the technique that may lead us there.

Still we may ask, "So what?" What in practice do these prophets have to tell us that could possibly be of use? Well, if it were the case, as the skeptics suggest, that they all arrived at different conclusions, then the answer, surely, would be "Very little." If one prophet said we should love our neighbors, and another said that all neighbors are bad news and should be done away with with all possible speed, then we could reasonably conclude that prophets were of no more use than the denizens of any barroom.

But in fact they do not. As the great nineteenth-century Hindu prophet Ramakrishna observed, all the great moral teachings of all the great prophets of all the great religions can reasonably be summarized in three edicts, all of which describe not particular actions that we should take but attitudes that we should seek to cultivate. These are personal humility, respect for fellow sentient beings, and reverence for the Universe as a whole. I have modified the wording somewhat, since Ramakrishna talked of respect for other people rather than for "fellow sentient beings" and of God rather than of the "Universe." But the essence is the same, slightly broadened and secularized.

I suggest that most people who think seriously about ethics, from whatever angle, would find it quite difficult to sustain serious objections to these propositions. Of course, they could take quite different stances. They might suggest, for example (as Nietzsche seemed to do), that personal arrogance and self-belief are worth more than wishy-washy personal humility, that it is foolish to care about anybody except oneself and one's immediate family, that nonhuman creatures are beneath consideration, and that the Universe as a whole—or at least the bit we are in contact with, namely Earth—is merely raw material for us to treat as we will. Such recommendations *might* be classed as "ethics," since they would, after all, be statements of how we ought to behave. But I suggest that most people who think about ethics would feel that such recommendations are bad. Ethics, most people feel, is about behaving *un*selfishly. Hitler thought he was a splendid fellow, but most of the rest of the world seems to disagree. The plea for personal humility, respect for fellows, and reverence for the Universe as a whole does strike a chord with most people; it does seem to provide a reasonable summary of what most of us feel is "good." As a grand overview, reduced to the simplest possible statement, I suggest it is difficult to improve on.

So where is the authority for this summary of ethics? If the prophets are truly inspired by God, and God is indeed in charge, then they have a right to speak out. But if there is no God, then what right do they have? And why should we take notice of them?

Perhaps, in reality, they have no "right." Perhaps the commandments of Moses were simply his own invention, and perhaps he had no more right to publish them than, say, Schubert had to compose his lieder or Shakespeare to write *Hamlet.* No more right, but no less, either. We, collective humanity who form their audience, have a perfect "right" to reject all of their inventions. We didn't ask Moses to tell us what our attitudes should be, or Shakespeare to trouble us with his plays. On the other hand, we are here, we have to get through life, and we cannot survive without the thoughts of other people. It is foolish to reject everything that everybody else says. It is sensible to accept at least some contributions as worthwhile. Given that we are trying to get through life in the company of other human beings, and indeed that we rely on the good will of other human beings, we might accept the proposals of Moses simply on pragmatic grounds. His authority springs simply from the fact that his comments seem helpful. (In short, even if we rejected the idea that Moses' or Christ's proposals do represent *absolute* morality, we might still accept those proposals on consequentialist grounds. Society would surely be better—kinder, safer, happier—if we acted in accordance with them.)

These three basic attitudes—personal humility, respect for fellow sentient creatures (human and otherwise), and reverence for the Universe as a whole—seem to me to take us a very long way in deciding which biotechnologies should be developed, and which we should at least soft-pedal.

For example, the idea that we should engineer breeds of cows, say, that can produce 4,000 gallons per year, or pigs that can breed like termites, seems to falter precisely because it shows no respect for the animals. It reduces them to a sac of milk or a womb. The prospect of a "designer child" with an IQ of 200 and aggression to match, eager and able to outsmart his or her fellows in the courtroom or stock exchange, falters on grounds both of personal humility and of respect. Is such personal success, achieved at others' expense, really so desirable? It may seem so in this present age, to people raised amid the politics of the past 25 years, but the politics of the past 25 years need not be the model for humankind.

More broadly, the notions of personal humility and respect for fellow creatures seem to lead naturally to the notion of "tools for conviviality," framed by the philosopher Ivan Illich in the 1960s. At that time it was fashionable to judge the goodness or badness of technologies according to their perceived sophistication. "Low" technologies (relatively simple devices such as windmills and bicycles) were typically judged to be good and appropriate for poor countries, while "high" technologies (the brainchildren of science, such as electronics) were supposed to be more appropriate for the rich.

Illich drew the lines differently. He suggested that a technology might be judged good if it increased the autonomy (independence, self-determination) of the users. These he called convivial. Examples did indeed include the bicycle, which improves personal mobility, and is largely under the owner's control, not least because bikes are relatively easy to repair; but they also included the telephone, which enables individuals to talk to individuals. Technologies were bad, by contrast, if they served to increase the power of one particular group of people—sometimes the creators of the technology, sometimes governments—over other people. Illich suggested that public broadcasting, at least as it was manifest in the sixties, when it was typically controlled by a single radio station, was of this type: run by a minority to influence the majority. Note that the telephone is judged good even though it is high-tech; and in fact bicycles that truly work are also high-tech (since they require modern tires and metal alloys).

The principle of conviviality seems entirely appropriate in judging the value of GMOs. If Indian peasant farmers had access to pigeon peas that had been engineered with genes that increased resistance to mildew, then this, on the face of things, seems a good thing. At least it would help those farmers continue with their traditional way of life. Of course it would not be so good if the suppliers of the peas also demanded some degree of fealty as a quid pro quo. If the peas were truly to enhance the farmers' autonomy and so pass the Illich test, the farmers would need, somehow or other, to control the technology that produced them. This is feasible. Genetic engineering, in the future, need not be a horrendously expensive technology and might well be carried out at a regional level, conceivably even at a village level, with local goals in mind.

By contrast, the wave of engineered cereals and rapeseed that have caused such a furor in Britain in recent years do *not* satisfy the Illich criterion of conviviality. They are produced by a few companies to reinforce a system of agriculture that is increasingly monocultural, reducing the range of crops grown and available to consumers, reducing the farmers to subcontractors, and focusing profits in fewer and fewer hands. Clearly, they are being deployed as agents of control. Protesters who dug up the engineered crops in Britain gave a variety of reasons for doing so. Some of those reasons do not seem to stand up to close scrutiny. Some (as the defenders of the crops were wont to emphasize) were simply Luddite, reflecting a general objection to technology. But the political objection was, and is, undoubtedly valid. The new crops, in the context of late-twentieth-century Britain, were part of a broad strategy to make agriculture even more industrial, and to bring it even more into the control of big business. That was widely perceived to be a bad thing. Again, the central objection is not that industry per se is bad or that capitalism is innately evil. But agriculture is important to all of us in many different ways, and it is not desirable to place control of it in the hands of a few people. It is an important principle of democracy that power should be spread as widely as practicable. Behind that lies the desire to protect personal autonomy; behind that, I suggest, lies respect for fellow human beings. In short, GMOs *could* be convivial, in Illich's sense. But the GMOs destined for British fields in the late twentieth century were not designed with conviviality in mind, and people were right to object to them (albeit often for the wrong reasons).

Finally, but crucially, we may observe that the scientists and companies who change the way crops are produced—by genetic engineering rather than by conventional breeding—have no mandate to do so. They have argued that they do not need one, since genetic engineering is merely an extension of conventional breeding. But that simply is not the case. The qualitative distinction is clear. In a democracy, anyone who does anything that affects the lives of the community as a whole needs a mandate to do so; at least, if "democracy" means anything at all.

But does the objection to GMOs—or to designer babies who become supercharged billionaires—spring only from the principles of personal humility and respect for others? What of the objection raised by Prince Charles, that this is "God's territory"? If we reject the idea of God, as is common in this secular age, does such an objection make any sense?

I suggest that it does. Again, I feel that we would do well to be cautious on absolutist grounds, to cultivate the feeling that such interventions really are beyond the pale. But this point can be made simply by consequentialist arguments.

Consider the point made earlier: that we can *never*, for inescapable reasons of logic, understand biological systems exhaustively. We may develop wonderful theories that seem to explain the observed facts beautifully, and indeed, we already have many such theories. But we can never know that we have taken all possible factors into account. Bridges and high-rise buildings still fall down, even though the technologies have been worked out over 10,000 years (at least), and the basic (Newtonian) physics has been developed intensively over 300 years. Still nature springs surprises. *Living* nature is much more complicated than bricks and steel and surely has many more surprises in store.

I suggested in chapters 8 and 9 that if the gene is compared to a word in a language, then the genome as a whole should be compared to literature, and I asked whether any of us would risk editing some ancient, sacred text in a foreign language if all we had was a dictionary. It would be reasonable to correct obvious blemishes, to fill in missing letters destroyed by foxing, for example. But to recast the text, to change the structure of sentences and the meanings implicit in them, would be presumptuous indeed.

But if we undertook to edit the text of the human genome, as opposed merely to correcting unmistakable blemishes, then the word *presumptuous* would not do. The appropriate word would be *hubris*. The word is overused, as so many words are these days. Soccer players who attempt to tackle one defender too many are accused of "hubris." But for the ancient Greeks, who invented the term, *hubris* had a truly chilling quality. It implied that a human being was trying to usurp the power of the gods. All human fortune, according to Greek religion, was entirely dependent on the will and whim of the gods. Success could be achieved only with their help and approval. If ever a Greek hero, flushed with some victory, supposed for an instant that his success sprang simply from his own courage and ability, he was sure to be cut down. To suppose that human beings are truly in charge was hubris, and this was instantly and ruthlessly punishable. Hubris, in short, is somewhat different from the Jewish concept of "blasphemy"—an offense against God—but it carries similar weight.

Many people have tried to list the kinds of things that might go wrong when we add genes to plants and grow them in fields in contact with other plants, or if we were to try to rearrange and enhance the genomes of human beings. Some of the suggested possibilities seem fairly chilling, although the most horrendous generally seem the most unlikely. The simple truth is, though, that we *just don't know,* and however much research is done, over however many decades or centuries, that will always be the case. We will never be able to anticipate all the tricks nature might play. To suppose otherwise, an ancient Greek would say, is hubris.

Ah, say the technophiles, but if all our ancestors had taken that attitude, then we would still be living in caves! So we might, except that we might not have made it as far as the caves. We might still be in the trees. But we can (in consequentialist rather than absolutist vein) do a cost-benefit analysis. What is to be gained by growing GMOs? If it helped human beings to continue living as they choose to live—for example, as farmers of pigeon peas in difficult environments—the gains would be quite large. The technology then would be "convivial." But if GMOs simply allow big companies to take over the world's agriculture, the gains would be at best equivocal. The known risks, that the genes might escape into wild plants and upset the local ecology, must be balanced against the putative gains, though we would have to suspend judgment on the unknowable risks.

What are the possible gains in producing designer babies? Again, we might spell them out in a Latin square: who is involved, who gains, who loses. The gains, I suggest, seem entirely selfish, and entirely concerned with the personal enrichment of the genetically manipulated offspring and the aggrandizement of his or her parents. The risks are horrendous. The ones we can guess are serious enough. As for the unknowable ones, we just have to envisage some equivalent, in a human context, of a bridge falling into a river—what may happen when technology runs ahead of understanding.

For some, though, the possible gains are tempting. What can outweigh the temptation? Nothing less, I suggest, than the moral conviction that gratuitous manipulation of the human genome is *wrong*. Consequentialist arguments based on benefit versus risk do not carry the necessary weight. We need the absolutist concept of hubris. This is, of course, a religious concept. It is an emotional response, carefully defined and deliberately cultivated. The concept of hubris (like that of blasphemy) works most easily if we also believe that the world is literally run by God, or by gods. But the feeling that gives rise to it must be cultivated anyway, simply as a matter of policy. Nothing less will do.

The concept of hubris, I suggest, may in turn be seen as a manifestation of the ethical principle defined by Ramakrishna that we should treat the Universe as a whole with respect. *Universe* can be seen as a secular translation of *God,* or *the gods.* The point, in both cases, is that Earth and its creatures—including our own physical and mental selves—are not simply commodities for us to manipulate at whim. By the same token, we should not need to ask whether the conservation of wild creatures is a good thing, or worthwhile. Of course it is. If our attitude to the world that we did not make is one of reverence, if we feel it is a privilege to live in such a world and have wild creatures around us, then how could it be otherwise?

In short, I suggest that very simple ethical principles—simple in the sense of fundamental—carry us a very long way in contemplating the uses and abuses of modern technologies. These principles are not the creations of lawyers, politicians, or even moral philosophers, but of prophets: people who operate at the level of the emotions, which (as David Hume said) must underpin all ethical deliberation. We need to return to these fundamentals, and to reiterate them. They will not provide us with off-the-shelf prescriptions in every situation. But they will provide broad guidelines, and without these we are lost. Indeed, since the new technologies could—if we simply stood back—redesign the human body and the human mind, then the human species as we now conceive it would be lost. There may be no absolute criteria of goodness or badness. There may be no literal God, and Kant's ambition to find such criteria without a God may have been forlorn. Even in a secular age, however, we need to pursue the notion that there *are* absolutes, and to cultivate again the arts of thinking in absolute terms.

The task may not be as difficult or arbitrary as it may seem. We should look again at the naturalistic fallacy. Perhaps, after all, the absolutes we seek may already lie within ourselves. Perhaps, indeed, like the rest of us, they are evolved.

THE STRENGTH OF HUMAN NATURE

The naturalistic fallacy does apply. It is naive simply to propose that what comes naturally is "right." It is *not* the case that it is good Darwinian policy for a man to rape, but if it were, that would still not justify rape. A woman may well serve her own reproductive interests most efficiently if she becomes pregnant by one kind of man (basically, one who is good at spreading his own genes) but allows another kind (a good, unselfish father) to bring up her children. Yet we tend to feel that the deception that would be involved is unsavory. Our own brains, our own consciences, tell us that what may seem most expedient for our genes at any one time may not be *right* after all.

What critics of evolutionary psychology commonly suppose to be "good" Darwinian policy generally turns out to be very crude Darwinian policy—the first approximation, the strategy that seems to emerge from the simplest kind of arithmetic. Darwinian policies have evolved by natural selection, and natural selection, in general, is far more subtle than many commentators seem to suppose. So we do indeed hate the coercion and the cruelty of rape, but the revulsion that we feel can itself be seen as an evolved response. We do not approve of the deception, when a woman cuckolds her husband, but again, evolutionary theorists (notably, in this context, John Tooby and Leda Cosmides) point out that human beings—and indeed all animals, and plants for that matter—have evolved intricate mechanisms for detecting deceptions of all kinds. More broadly, Robert Frank of Cornell University has argued (especially in *Passions Within Reason: The Strategic Role of the Emotions,* W. W. Norton, New York, 1988) that human beings have *evolved* a huge range of subtle emotions and behaviors that have to do with dignity, honor, and trust, as well as with the crude passions of anger and fear. David Hume himself distinguished between first-order passions (such as anger and fear) and second-order emotions (such as respect and trust). Most philosophers have tended to assume that the second-order emotions—often seen as "higher" emotions—are peculiar to human beings, and are yet another diagnostic feature of humanity. But modern scholars of animal behavior perceive that animals, too, recognize concepts such as respect and dignity. These emotional refinements, in fact, run very deep indeed, and are themselves evolved.

But then, if we put all our prejudices aside—the prejudice that says human beings are qualitatively different from all other creatures, and the prejudice that says complex behaviors and emotions *cannot* be underpinned by genes—we would find ourselves asking, "How could it not be so?" No one doubts, after all, that human beings are, by instinct, social creatures. Prolonged solitude in all literature is considered to be unfortunate, heroic, or weird, but it is never considered "normal." As social creatures we need a repertoire of behavior and emotional responses of enormous range and subtlety. Think how many social nuances are involved in the simplest day-to-day transactions, like ordering a meal in a restaurant. No one supposes in these enlightened days that the waiters are literally inferior to the clients, everyone present may feel they want to relax and have a good time, and most of the participants are strangers to each other and often do not share a common first language. Even in such a simple scenario, the roles that each participant must play, if all are to relate satisfactorily to all the others, have endless ramifications. Yet on most days, the participants in most restaurants around the world play their roles perfectly. The waiters are deferential but not servile, the customers are in charge but not aggressive, and people relax and laugh, but if they overdo it they are condemned by staff and fellow guests alike as oafs. A computer that could play such roles so flawlessly would require a much larger program than the whole of NASA is liable to need for the rest of its existence. Yet most of us breeze effortlessly into restaurants, the classroom, stores, and any of a variety of social situations and behave impeccably (or at least well enough) without thinking about it. Indeed, if we *did* think about it, we would screw it up. We would become self-conscious, thinking about everyday behaviors that ought to be innate. Once we become self-conscious, we cease to be "natural" human beings and become bad actors, unable even to sit down on a chair without fear of disaster.

We can behave as social beings only *because* we are so beautifully programmed, which in the end means genetically programmed, to do so. Of course we learn different manners in different societies, and behave differently in different circumstances. But our conscious minds, and our learned manners, would be hopelessly at sea if they could not draw on the deep well of innate responses that have evolved not simply over the last 5 million years of specifically human evolution but through all the hundreds of millions of years in which our prehuman ancestors honed the basic social responses.

Psychologists and computer scientists now agree that emotions—which, crucially, embrace the more subtle responses of appreciativeness, kindness, respect, and all the other perceived human virtues—play a vital role in framing those social responses. Computer scientists see emotions as a kind of shorthand. They are not detailed programs, telling the brain to carry out strings of particular actions. They are general instructions, setting the tone and drift of the brain's activities. Computer scientists now build artificial emotions into robots, because if they do not, the robots cannot make decisions. If the robots simply use their brains, then they spend all their time weighing pros and cons, and in real life, there are too many factors to take into account. Without emotions we are paralyzed.

In short, we really should respect the idea of the naturalistic fallacy: we really should not assume that what comes naturally to us is necessarily, ipso facto "right." On the other hand, we need not assume that what is "natural" is necessarily brutish and crude. G. E. Moore was writing at the beginning of the twentieth century, when nature was still perceived to be "red in tooth and claw." At that time, the "natural" emotions were felt to be those of the zoo, and a primitive zoo at that, where the animals are locked up, deprived, and behaving pathologically. Aggression and naked fear were felt to be "natural," while the fruits of human kindness—benignity, respect for others' dignity—were felt to be exclusively human refinements and, indeed, the stuff of civilization. Most people still seem to assume that this is the case, that "nature" is innately crude and only the human brain is subtle, and only after a few centuries of civilization. This, I suggest, is the prevailing human prejudice. But it just ain't so, and a little reflection shows that it *cannot* be so. If human beings—or any social animal—behaved as crudely as Tennyson's famous phrase suggested, then no creature could, in practice, live socially. Redness in tooth and claw simply will not do. Social interactions are *bound* to be more complex than that. All social animals are socially subtle. Human beings might reasonably claim to be the most subtle of all, for all kinds of reasons. But the subtlety is not simply the creation of our own vast brains. Our brains build on, and make manifest, the social subtlety that was already well established in our most ancient ancestors, and that social subtlety includes the repertoire of the most delicate emotions that have ever been acknowledged by the world's poets.

So yes, we should be prepared to use our brains to override "nature." We should seek to ensure that our brains, or at least our "minds," are the ultimate arbiters. But we should not underestimate nature. Our inherited nature *includes* much of what any of us would call morality; it *includes* a respect for fellow creatures. So although we might override nature, we would do well to listen to nature, too. We all know what a "conscience" is: the "inner voice" that tells us we are behaving badly. We need not doubt that that "inner voice" is itself evolved, calling to us from our difficult days on the African plains.

Indeed, I am inclined to suggest that the prime trick of prophets, whom I have extolled, is not to tune in to the voice of God, as they themselves have generally believed or at least proclaimed to be the case. In reality, they are listening to their own inner voice—to evolved nature. They need tranquillity to do this. Our evolved nature, shaped by natural selection, is not trivial and it is not crude. It is worth listening to.

How do such arguments help us in weighing the new technologies? Well, some philosophers of my acquaintance take an entirely "rational" line. For them, the Latin square approach to morality is enough. In the issue of cloning, for example, they are content to list the people involved along one axis and the pros and cons along the other, add in a factor X to acknowledge the risks that are yet unknown, and then add up the figures. If the perceived pros seem to outweigh the perceived cons, then (say the ultrarationalists) cloning should be seen to be acceptable.

Others, though—I am inclined to say "most people," certainly most of those I have talked to—have additional misgivings that have nothing to do with quantifiable advantages and disadvantages. Many people simply feel uneasy about cloning. It just doesn't seem *right*. The rationalists reply that *all* technologies seem awkward when they are new; it's just a question of getting used to them. Many people objected to IVF when it was first introduced even more stridently than many now object to cloning, yet IVF has become an accepted reproductive treatment. Once people see the advantages of cloning (the rationalists suggest), they will also realize the error of their ways and welcome the new advance. The same will surely apply to the designer baby. In a few hundred years, when the technologies are running smoothly, no sensible person will see any objection at all. Those who do object will very properly be sidelined, the flat-earthers of the twenty-third century.

I would like to suggest that we should listen to our misgivings and strive (as the prophets do) to make them explicit. These misgivings are intuitive. They are built into us. They are evolved responses, and we should not take our own evolution lightly. Evolution does not necessarily lead to crude results. We should not assume that the thought processes that we call rationality are necessarily superior. Furthermore, the emotional responses that are built into us in a sense *are* us. We are evolved creatures, and these evolved emotions are part of the evolved person. Of course we should apply our rational brains when making moral decisions, for the alternative is simply to shoot from the hip. But we should not assume that our intuitive responses are merely "primitive," or that because they are primitive we can disregard them. Our primitive emotions are a very important part of *us,* what it means to be a human being, and if we override those intuitions lightly, then we might as well hand over our lives to computers. Some people evidently would like to do that. Maybe if we did so, we could run our human affairs more efficiently. It's a big "maybe," but in any case, if we did so, our affairs would cease to be "human" at all.

So let's return to the ancient and deep concept of hubris. In the manner of evolutionary psychology, we might propose that this concept is itself an evolved response and ask how and why it evolved, what it represented to our early ancestors, in whom it first presumably arose. I suggest that the root of hubris is dread, and the cause of the dread is, simply enough, the unknown. Every creature—human and otherwise—depends for survival on knowledge of its environment. It has to *know* (or at least be as sure as is possible) what the possibilities and the dangers are, and what outcome will result from any particular action. A rabbit needs to know where the burrows are so it can dash for cover when the fox appears; it needs to know, too, that the burrow does not contain a snake, which would be just as bad as the fox. If a rabbit is placed in a strange territory, and does not know where the holes are or whether they are safe, you can see its terror. Animals in novel environments first take time to explore them. All animals need to know, or feel, that they are in control.

Dread, I suggest, is the feeling that human beings (and probably other animals) get when they are in unknown territory, when they do not know the dangers or how to respond sensibly to any danger that might arise. Modern human beings seem to have evolved in the open woods and plains of southern and eastern Africa, and although many modern people have returned to the forests (which, after all, occupy most of the tropics), many of us have a dread of forests. Forests contribute much to the terrors of gothic literature. The very word *jungle* excites a frisson, with its implied confusions and hidden threats. *Jungle warfare* is innately terrifying.

We have all felt the dread that comes from circumstances we do not feel we understand and cannot control: lost in a forest, stuck up a cliff, out at sea. In such circumstances one thought is paramount: "Get out of here!" We may have to abandon all our gear, cameras and wallets and picnic baskets, and be forced to sustain injury. But the priority is clear—get out, because events are overtaking you.

Hubris surely is the formal, poetic expression of that primitive dread, or rather, the directive intended to prevent people getting into dreadful situations in the first place. The Greeks were wonderfully in control of their environment; none had ever been more so. They had excellent technologies and stable societies supported by clear formal laws and moral philosophy. Yet they were also subject to famine, earthquake, invasion, and infection—instant, random death from a hundred quarters. Archaeology and written history record several devastating epidemics, including both malaria and plague. The Greeks knew, then, that for all their rationality and confidence, they were not *really* in charge of the world around them. They were in charge only up to a point, of the bits that immediately concerned them, at least temporarily. They knew full well that they should take no liberties. They had very clear ideas of where it was safe to live and who it was safe to do battle with, and, vitally, what behaviors were acceptable and what were not. Those who exceeded the bounds were always liable to be struck down. The concept of hubris kept them on track.

In the context of the modern, very powerful biotechnologies, I think we are absolutely right to invoke the concept of hubris, and to take it seriously. We *know* we cannot understand the world absolutely, and the more we seek to manipulate living systems, the more we are likely to be taken nastily by surprise. The modern biotechnologies in their extreme form really do present us with a "no-win situation," as the modern jargon has it. For if we take the new technologies to their ultimate conclusion (and if, by good luck, we fail to encounter disaster along the way), then we will end up redesigning humanity. Human beings, *as we know them,* will be superseded. Whatever the qualities of the new race that may come after us, that seems to me very like a species of suicide. There is a lot wrong with humanity. But we are not trivial creatures; and the fact that we have *evolved* to be as we are, far from a being a drawback, is in fact one of our greatest assets, for natural selection over long periods can operate very subtly, and produces subtle creatures. The possible gains that biotechnologists might make by changing our genes are entirely hypothetical and the risks are clear (or unknowable, which is worse). We might well try to avoid the ill effects of the more obvious blemishes—the frank pathologies like Tay-Sachs disease that lead to such distress—but to attempt a wholesale editing of the human genome is simply not sensible. More than that: such an attempt would misconceive our abilities. It would indeed be hubristic. We should accept ourselves as we are, warts and all. Beneath the warts, after all, there is much to be pleased with.

The belief that this is so springs from an emotional response, and the emotional response can be seen to have primitive roots, nothing more nor less than the dread of the unknown. It would be misguided to ignore our instincts. We have reason to suppose that our instincts are good, precisely *because* they are evolved, hammered out over many thousands of years by natural selection and "designed" to help us survive in the rough, tough world as social creatures, alongside others of our own kind. It would be hard indeed to improve on that inheritance. Besides, our instincts are *us,* and if we betray them, we betray ourselves.

Suppose, then, that some readers agree with this analysis, or at least with its conclusion: that we should leave well enough alone and seek simply to make life better for ourselves, each other, and other creatures while we are on this Earth. What are the chances that such fine sentiments will be acted on? In practice, are we human beings truly in command of our own technologies? Or are they running away from us, driven by forces that are already beyond our control?

WHAT CAN CONTROL THE NEW TECHNOLOGIES?

Lee Silver, professor of biology at Princeton University and author of *Remaking Eden* (Weidenfeld & Nicolson, London, 1998), suggests that human beings in the future *will* practice cloning, and *will* manipulate the genomes of their offspring to the point where they will, in effect, produce a new stratum of society. These will be the "Genrich" people, as opposed to the rest of us who are content (or stuck) with the genes we inherited by normal means. Indeed, he suggests that the Genrich might eventually evolve into a new species: able to mate successfully with each other but not with unmanipulated hoi polloi. However, the biological technicalities may be somewhat questionable. A young lad fitted with genes that confer some outstanding quality (such as speed of thought) might be well advised, genetically, to find himself a wife whose genome has not been tampered with. Among other things, the offspring of such a union might benefit from the resulting hybrid vigor. Be that as it may, Silver presents a crucial issue with admirable clarity— not *is it right* to clone people, or to produce designer babies, but *will these things happen,* whether or not we think they are "right"?

Silver has no doubt that they will. What will drive these biotechnologies, he suggests, are market forces; market forces cannot, in the long term, be gainsaid. There is both pull and push: demand on the one hand, and an eagerness to satisfy demand on the other. There is no stronger emotion than the desire to reproduce, and people seem, at least in general, to prefer offspring who are genetically related to themselves. One in eight couples suffers from some reproductive problems (and there is news, early in 2000, that the figure might be rising closer to one in six). Many have already shown their willingness to undergo and to pay for exotic and time-consuming technologies. By the late 1990s an estimated 150,000 children had been born by IVF, and by 2005 there could be half a million. In some cases, the cloning of one or the other parent will seem to offer the only means to produce a genetically related baby. Besides, in this flexible age, same-sex couples are becoming commonplace, and for them or for parents who really want to remain single, cloning seems to offer the only feasible option. Cloning is more exotic than IVF, but as the decades pass the technology will become more-or-less routine. At that time, the demand will *of course* be enormous. People will be prepared to pay, too. Already they may pay $50,000 for IVF, without guarantee of success. Cloning (when the problems are ironed out) should be more surefire, and the fees could well be even higher. People will pay.

Similarly, parents these days commonly spend $100,000 on their children's university education. Would they not spend another $20,000 on a few genes that would enable their offspring to make better use of that education? Go to Harvard with an IQ of 130 and you should do well. Start with an IQ of 140 and you should do even better. That is the logic. Where's the harm?

There will be no shortage of willing suppliers, either. Even if we personally may disapprove of cloning or designer babies, we should not assume that those who might supply such services are necessarily cynical, or evil. Many do not disapprove. Many do not find the objections convincing. The case made above, for conservatism and acknowledgment of hubris, has yet to be widely tested in the public arena. Besides, many clinicians argue that it is simply not their job to judge the requests of their patients, patients whom they see, very reasonably, as clients. It is their job as physicians to reduce human suffering. If people are suffering because their reproductive instincts are frustrated (and no human suffering is more keenly felt), then they should do what they can to help. If cloning seems the most feasible option, then so be it. Furthermore, those who supply such services know that they can command high fees; indeed, a successful clinic might offer a legitimate route to fortune. Financial return may not be the prime motive for a physician, but it helps.

Silver is right to suggest that market forces are tremendously powerful, and no more so than in the field of human reproduction or (perceived) human improvement. Furthermore, in the present world climate, the free market is seen to be innately, morally, desirable. It is the natural expression of democracy. It supplies what people say they want. That is its raison d'être and its sole route to survival. Furthermore, the collapse of centralized economies in the latter half of the twentieth century seems to demonstrate that of all the economic systems devised so far, *only* the free market can efficiently supply what people want. If the things that people want do no harm to third parties—if the market does not trade in weapons of mass destruction, for example—then how can the market be criticized? What is better, in theory or in practice?

The moral defense of the free market runs even deeper than this. The United States and western Europe have long seen themselves as the world's proponents and defenders of democracy, and so indeed they are, if anyone is. Yet there are many interpretations of "democracy." Britain seems to stress majority rule and, in these multicultural days, paying significant attention to the special demands of minorities. On the whole, if 60 percent of the people say they want such and such a thing, then the remaining 40 percent, albeit with reluctance, accepts that this is the way society is and puts up with it. The United States sets more store by the democratic principle of personal freedom: some of the politicians who advocated America's secession from Britain in 1776 specifically condemned what they saw as the oppressive principle of majority rule. The *point* of creating their brand-new country, they felt—this brave new world—was to enable *men* (the eighteenth-century word for "people") to do their own thing. If they merely kowtowed to the majority, then they were no better off than they would be in a quasi-feudal society ruled by aristocrats. Personal freedom remains the battle cry of modern American Republicanism. The free market is seen as the expression of that freedom, which on the face of things is a reasonable point of view. On such grounds, good American Republicans, or indeed liberals, could perfectly well defend the right of others to indulge in cloning, or to create designer babies, even if they themselves disapproved of such procedures. Thus in America in particular, the free market is very well protected by layers and layers of moral philosophy, history, and law. Such an institution cannot easily be overridden; neither (many would say) should it be.

So is that the end of the matter? We might pontificate as moral philosophers, and even appeal to religion as I have done, albeit in a nonatheistic form, but what can anybody say that can divert the market? What *should* they say? But if moral philosophy and religious conviction, and the laws that may derive from them, cannot in practice override the market, what can? Does this not mean that *whatever* misgivings some of us may have, the market will drive inexorably forward? The time scale must also be taken into consideration. The market might lose the first few rounds of discussion, over the next few decades, but the technologies will still be around, steadily becoming more and more efficient, in 200 years' time, or 500, or 1,000. And the market will still be there. Why don't we just give up now and lie back and enjoy the changes that are bound to come about?

But the "battle" is not yet lost. Even if it is lost eventually, we should go down fighting. It is difficult to doubt, after the history of the twentieth century, that the free market does have advantages, and it is surely foolish at this stage of history to mount any political revolutions.

Does the market really reflect the will of the people, or most of the people, all of the time or most of the time? For all its apparent power and precision, the free market is a limited and blunt instrument. It supplies commodities to people who can pay. The richer they are, the keener it is to meet their desires. It contains no innate mechanisms of social justice. There is nothing in market forces to cater to those who cannot pay at all, to bring them up from the bottom. If anything, the free market left to itself reinforces *in*justice. The rich grow richer and the poor grow poorer. A society that dislikes poverty and injustice has to modify the free market. In all civilized societies, for this and other reasons, markets have to be constrained by laws or at least by acknowledged codes of practice.

Nobody seriously doubts this. The argument is primarily about the *degree* of control that society at large should exercise over the market, and, in these days of the European Community and the World Trade Organization, about what "society" actually means. Is the "society" the nation, as was once taken to be self-evident, or is it the entire quasi-federated continent, or indeed the whole world? These points are highly pertinent, but we can put them to one side for the moment. What matters is the general principle. The free market is here to stay, and we should probably be grateful for that, but the market should be the instrument of society and must not itself dictate the shape and mores of the society.

What, though, in a democracy—especially one committed to personal freedom—can prevent the supremacy of the market? Who or what has the moral right to override it?

Well, people at large, of course, have the moral right to constrain the market, to tell it what it can do and what it cannot. The means by which people at large exercise that right are twofold: through the law, and through the market itself. Despite the apparent attractions of cloning and designer babies, people may decide, after all, that they just do not want what is on offer. A few people might welcome the new technologies, just as a few opt to have their breasts reshaped or their lips made plumper. The law alone could not prevent such minorities from indulging their desires, for whatever is made illegal in the United States or western Europe could become a cottage industry on some far-flung island of no particular affiliation. But if cloning and genetic manipulation remain minority pursuits—perhaps somewhat looked down on, considered infra dig—then they will not dramatically affect the evolution of the human species. They will still be important, since they would affect individuals, but their importance would not, as has sometimes been suggested, be of seismic significance.

How we make laws to constrain the market, and what should be in those laws, of course raises all the issues argued in the first part of this chapter. I have suggested that whatever we do, we must dig deep, invoking and leaning heavily on our inherited emotions.

But if we do that, we might find that law is to a large extent superfluous. People might yet decide, as it were spontaneously, that they simply do not want what these new biotechnologies seem to offer. Some exotic procedures—like IVF—have caught on; initial revulsion over such technologies has tended to give way to acceptance. But people in various societies have also shown that they can and do *reject* particular technologies simply because they feel uneasy with them. Somehow, for whatever reason, the technologies seem threatening. Advocates of those technologies invariably argue that whoever rejects them is simply perverse, or "ignorant"; such discussions quickly become acrimonious. I am inclined to suggest, though, that our instincts can be far more sensible than the ostensible "rationality" of the advocates. The latter, after all, are pitting their professional knowledge against 5 million years of evolved wisdom. Thus many people have rejected nuclear power and some societies have banned it; although the advocates proclaim its safety (even after Chernobyl and various near misses), the widespread rejection seems eminently sensible. In Britain, too, the high-rises that the ultrarationalist planners saw as the answer to the postwar housing shortage are now being knocked down. For families, at least, they are a disaster, as the families themselves said at the outset would be the case. In short, technologies may come into being with tremendous piles of cash and intellectual blackmail behind them and yet be thrown out. If they don't appeal, they don't appeal.

Might cloning and designer babies suffer such a fate? It is too early to say, of course, but they could. People's enthusiasm for present-day reproductive technologies is not as unalloyed as their advocates sometimes claim. Many people seem to have taken to IVF. However, although IVF is an exotic technology, it is usually carried out in the context of marriage. It is usually intended to help married men and women, or at least serious partners, to have children together. Thus, while the technology itself is exotic, it is aimed (in effect) to reinforce the most homely and conventional of all human institutions: the heterosexual, monogamous partnership that leads to the nuclear family. In context, IVF is cozy.

By contrast, the alleged promise of artificial insemination (AI) seems largely unrealized. AI was first used formally in human beings nearly 250 years ago, by the Scottish physician John Hunter. It was offered as a commercial service in the late nineteenth century. Freezing of human sperm became technically straightforward after the 1950s, so AI became logistically simple: the donor no longer had to be in the next room. With modern marketing (including the Internet), it is positively easy for virtually any woman to buy sperm from the most accomplished men in the world: with A's from Stanford in the subject of the mother's choice and gold medals in the Olympic games. Many women proclaim that they would rather live without the encumbrance of men. Many have suggested that they might live like lionesses, in prides, bringing up their children in nurseries, collectively cared for, and consorting with men (if at all) only for social purposes.

So why don't they? Why, instead, do so many intelligent, determined, and free-spirited women settle for an average couch potato, a clone of Homer Simpson? Perhaps—just perhaps—because society these past few hundred years has been based on monogamous man-woman partnerships, so that this way of life has become the norm. Perhaps too (so some feminists would claim) these monogamous relationships were invented by men, ultimately for the convenience of men. If ways of life are purely cultural, then surely they are subject to change. What is normal in one society, or set of circumstances, may be quite different in another. Perhaps as men lose their psychological and economic hold over women, more and more women will reject simple heterosexual monogamy in favor of the pride-of-lionesses scenario, or any other scenario they may care to adopt. Perhaps if all these things are the case, then reproduction by AI will become commonplace, and monogamous heterosexuality with the live-in male as genetic father will seem reactionary, or merely quaint.

But perhaps—this is only a "perhaps," but all possibilities ought to be considered if we want to arrive at the truth—women prefer to live monogamously with men, and to bear the children of the men they live with, because that is the way they are. Their preference for such a way of life may simply be evolved. If such a predilection were evolved, it would make perfect sense. Men on the whole (or males of any species) are not happy to expend resources on children who are not their own (or not so happy, at any rate); and women want partners to help them raise their children, and know that their partners are more likely to stay faithful (and helpful) when the children are their own. So there are deep evolutionary reasons for heterosexual monogamy. This does not mean that this is the only possible arrangement that human beings can feel comfortable with, for of course we are flexible creatures. Even if heterosexual monogamy was an evolved response and was in this sense "natural," this would not make it "right" in any absolute sense, or so says the naturalistic fallacy. It does mean, however, that heterosexual monogamy is what many people, men and women, feel most comfortable with; and what they feel comfortable with, they will practice. IVF, as presently carried out, generally reinforces heterosexual monogamy. AI—though technically much simpler—seems to threaten heterosexual monogamy. Life as a lioness manqué may have many theoretical advantages, but women, on the whole, don't really seem to want to live that way; the technology that could support such a way of life has not, in fact, caught on. Maybe it will one day. Maybe the news has not yet filtered through. But maybe, too, people will *always* prefer traditional family life and reject any perceived threats to it, simply because that's the way people are. "Rationalists" who proclaim there is a better way have to answer the question, "Why?"

In the same way that people *seem* largely to have eschewed the much-vaunted advantages of AI, so they might also continue to reject cloning and genetic manipulation. They may simply continue to feel that it goes against the grain, and if that is so, then all the pressures of the technophiles will not budge them. Of course, there will be a few cloned babies and "Genrich" children, just as there are a few all-women societies with babies raised in nurseries. But the idea that such ways of life will become the norm—that sex among humans will be purely for social reasons—is, we may reasonably guess, nonsense. My caveats of course are speculation, but I feel they are reasonable speculation. I suggest, too, that if people do feel more comfortable with heterosexual monogamy, and with babies produced exclusively by sex and without added genes, then it would be positively evil to persuade them otherwise. Whatever the naturalistic fallacy says, the natural instincts and predilections of human beings must be taken seriously. In a secular world it is hard to see what should be taken *more* seriously.

Taken all in all, then, there are grounds for cautious optimism. The fundamental ethical principles summarized by Ramakrishna—personal humility, respect for fellows, and reverence for the Universe as a whole—are simple yet seem robust. First, they consolidate many centuries of contemplation by serious seekers after truth, and second, they seem to summarize those aspects of our evolved instincts that have to do with social living. There is such a thing as human nature, and human nature is much more subtle and innately "moral" than nonbiologists (or even many biologists) choose to believe. The naturalistic fallacy does apply—we must use our brains to seek principles of ethics independently of our evolved instincts. Yet we would do well to explore and respect our instincts, and indeed we ignore them at our peril. For one thing, our instincts have evolved to help us to be social beings (teleologically speaking), and in general they have done a very good job. For another, our instincts are *us,* and if ethics are not about us, then what are they about? The free market is with us, is powerful, and at present seems hard to improve on. But it cannot be relied on to do what is "good" (unless, perversely, we define *good* as what the market is able to produce), so we must be prepared to control it and, if necessary, to override it. *Because* we are evolved creatures, however—evolved as social beings—our built-in instincts should help us to define what we want the market to do and to reject its excesses. We may well continue to feel that the more exotic promises of modern biotechnology *are* excesses. We *can* resist technophilia. To some extent, we have already shown an ability to do so.

The new technologies, taken to extremes, threaten the idea of humanity. We now need to ask as a matter of urgency who we really are and what we really value about ourselves. It could all be changed, after all—we ourselves could be changed—perhaps simply by commercial forces that we have allowed to drift beyond our control. If that is not serious, it is hard to see what is.

This book has focused on Gregor Mendel. I have tried to show how modern genetics, the obsession of modern humankind, flows so naturally from his gloriously simple but far from simplistic ideas. All that has followed since is footnotes. Mendel was born into a rustic, feudal society, but he was truly a modern scientist. He was also a cleric and the most benign of men: courteous, generous, socially conscious. Indeed, he has a great deal to teach us.

Sources and Further Reading

The following is intended as an informal reading list for those who want to get a little more deeply into the subject—and branch out from there.

Chapters 2 and 3

The definitive modern biography of Mendel is *Gregor Mendel: The First Geneticist,* by Vitezslav Orel (Oxford: Oxford University Press, 1996).

Chapter 4

An excellent general textbook is *Genetics,* by Peter J. Russell, 3d ed. (London: HarperCollins, 1992).

See also my own book *The Engineer in the Garden* (London: Jonathan Cape, 1993) and *The Language of the Genes,* by Steve Jones (London: HarperCollins, 1993). Matt Ridley's *Genome* (London: Fourth Estate, 1999) is also excellent.

Chapter 5

Molecular Biology of the Cell, edited by Bruce Alberts and others (New York: Garland Publishing, 1999), is excellent.

Everyone should read *The Double Helix,* by James Watson (New York: Atheneum, 1968). *The DNA Story,* by James D. Watson and John Tooze (San Francisco:

W. H. Freeman, 1981), is a highly enlightening and diverting compendium of papers and articles tracing the elucidation of DNA's structure and function.

Chapter 6

The seminal work is, of course, Charles Darwin's own *On the Origin of the Species by Means of Natural Selection,* originally published by John Murray in 1859. The Penguin Classics edition (Harmondsworth, U.K.: Penguin Books) dates from 1985; the quotation on heredity is on page 76.

Everyone should read *The Selfish Gene,* by Richard Dawkins (Oxford: Oxford University Press, 1976), which shows why it is necessary to shift the emphasis from selection of individuals to selection of genes.

Highly recommended modern treatments include *Evolution,* by Mark Ridley (Oxford: Blackwell Science, 1996); *Darwin's Dangerous Idea,* by Daniel C. Dennett (New York: Simon and Schuster, 1995); and *The Darwinian Paradigm,* by Michael Ruse (London: Routledge, 1989).

Excellent modern biographies are Janet Brown's *Charles Darwin* (London: Pimlico, 1996) and *Darwin,* by Adrian Desmond and James Moore (New York: Warner Books, 1992).

Chapter 7

The book that started the whole thing is E. O. Wilson's *Sociobiology: A New Synthesis* (Cambridge, Mass.: Harvard University Press, 1975).

For friendly introductions to the modern subject, see *Evolutionary Psychology: A Critical Introduction,* by Christopher Badcock (Cambridge: Polity Press, 2000); *Introducing Evolutionary Psychology,* by Dylan Evans and Oscar Zarate (New York: Totem Books USA, 2000); and *Human Nature After Darwin,* by Janet Radcliffe Richards (Buckingham: Open University, 1999).

For grand overviews, see *Evolutionary Psychology: An Introduction,* by Leda Cosmides and John Tooby (London: Weidenfeld and Nicolson, 1999), and *Handbook of Evolutionary Psychology,* edited by Charles Crawford and Dennis L. Krebs (Mahwah, N.J.: Lawrence Erlbaum Associates, 1998).

See also *Games, Sex, and Evolution,* by John Maynard Smith (Hemel Hempstead, U.K.: Harvester, 1988); Helena Cronin's *The Ant and the Peacock: Altruism and Sexual Selection from Darwin to Today* (Cambridge: Cambridge University Press, 1991); Matt Ridley's *The Red Queen* (Harmondsworth, U.K.: Penguin Books, 1994); Steve Pinker's *How the Mind Works* (Harmondsworth, U.K.: Penguin Books, 1998);

Geoffrey Miller's *The Mating Mind* (London: William Heinemann, 2000); and *The Truth About Cinderella: A Darwinian View of Parental Love,* by Martin Daly and Margo Wilson (London: Weidenfeld and Nicolson, 1998).

Chapter 8

See my own books, *Food Plants for the Future* (Oxford: Basil Blackwell, 1988) and *Last Animals at the Zoo* (London: Hutchinson, 1991).

Chapter 9

Books on modern biotechnologies applied to human beings include *Remaking Eden,* by Lee Silver (London: Weidenfeld and Nicolson, 1998), and *The Second Creation: Dolly and the Age of Biological Control,* by Ian Wilmut, Keith Campbell, and Colin Tudge (New York: Farrar, Straus and Giroux, 2000).

Chapter 10

I find *The Oxford Companion to Philosophy,* edited by Ted Honderich (Oxford: Oxford University Press, 1995), as good a way into moral (and other) philosophy as any.

77